PREPARATION OF THIS DOCUMENT

This document is the result of a series of lecture notes prepared by the author and utilized at different workshops and training activities organized by the FAO/DANIDA Project on Training on Fish Technology and Quality Control (GCP/INT/391/DEN) since 1986. Such activities, carried out in Africa, Asia, Latin America and the Caribbean provided an excellent opportunity to improve the text according to practical experience obtained in the field.

The author has also utilized part of the material presented as lecture notes at the Technological University, Lyngby, Denmark, at the Royal Veterinary and Agricultural University, Copenhagen, and at the Ålborg University, Ålborg, Denmark.

FAO has decided to publish it as an FAO Fisheries Technical Paper rather than as a project publication, to allow for a widespread diffusion in view of the worldwide relevance of the subject discussed.

The present document was originally prepared for use at courses on assurance of seafood quality conducted for trainees with a basic knowledge of food microbiology or biochemistry. However, for people with practical experience working on quality assurance in the fish industry the document provides the necessary background information and guidelines for their daily work.

Many people have provided constructive criticisms, useful suggestions and contributions, in particular Dr Susanne Knøchel and Professor Mogens Jakobsen, both at the Royal Veterinary and Agricultural University, Copenhagen, who have contributed with section 6.1 and sections 5.2 and 6.2 respectively. The document was produced and edited by Mr H. Lupin (FAO/FIIU, Project Manager of GCP/INT/391/DEN).

The bibliographical references have been presented as submitted by the author.

Distribution

FAO Fisheries Department
FAO Regional Fisheries Officers
Selector HP
FAO fisheries field projects
Author
DANIDA
Technological University, Lyngby, Denmark
Royal Veterinary and Agricultural University, Copenhagen, Denmark
Ålborg University, Ålborg, Denmark

Huss, H.H.
Assurance of seafood quality.
FAO Fisheries Technical Paper. No. 334. Rome, FAO. 1993. 169p.

ABSTRACT

This document is primarily focused on the application of the Hazard Analysis Critical Control Point (HACCP) system to the fish industry. The document reviews in detail the potential hazards related to public health and spoilage related to fish and fish products, and discuss the utilization of HACCP in different type of fish industries. It contains a chapter making clear the limitations of classical fish inspection and quality control methods based solely on the analysis of final samples. A brief introduction about the relationship between the HACCP system and the ISO 9000 series is also included. The document is completed with chapters related to cleaning and sanitation and establishments for seafood processing, primarily seen from the HACCP point of view.

ACKNOWLEDGEMENTS

The author is indebted to a great number of colleagues, FAO/DANIDA workshop participants and students, who have given constructive criticism and useful comments of early drafts.

Particular thanks are due to Dr. Lone Gram, senior researcher, at the Technological Laboratory, Danish Ministry of Fisheries, which enthusiasm, untiring effort and affection to detail and quality of work has been a great stimulation and drive to complete this document.

Special thanks are given To Dr. Susanne Knøchel, senior researcher and Professor Mogens Jakobsen, both at the Royal Veterinary and Agricultural University, Copenhagen. Their contribution of sections within their special expertise is greatly appreciated.

Mr. Karim Ben Embarek and Ms. Bettina Spanggaard, both Ph.D. students, have been both helpful in reading proofs of manuscripts and preparing index. Finally special recognition is given to Maria Henk and Inge Andersen of the Technological Laboratory, Danish Ministry of Fisheries, for skilful secretarial assistance in the preparation of the document.

Assurance of seafood quality

FAO FISHERIES TECHNICAL PAPER

334

by
H.H. Huss
Technological Laboratory
Ministry of Fisheries
Denmark

Food and Agriculture Organization of the United Nations

Rome, 1994

The designations employed and the presentation of material in this publication do not imply the expression of any opinion whatsoever on the part of the Food and Agriculture Organization of the United Nations concerning the legal status of any country, territory, city or area or of its authorities, or concerning the delimitation of its frontiers or boundaries.

M-40
ISBN 92-5-103446-X

FOREWORD

The Food and Agricultural Organization of the United Nations (FAO) has always recognized the need for quality assurance as an essential discipline to guarantee safe, wholesome and functional fisheries products.

No food production, processing, distribution company or organization can be self-sustained in the medium and long term unless the issues of quality, including the safety aspect, are properly recognized and addressed, and an appropriate quality system is put into operation in the processing establishment.

The practical limitations of classic methodologies of fish inspection and quality control, based on analysis of final samples, have been known for several years. This is why many Governments and the fishery industry in developed and developing countries have embarked on an important conceptual change of fish related regulations, including inspection, handling and processing, import-export and marketing.

The need for effective quality assurance systems is further underlined by the fact that global fish production has reached a plateau and that further increases in wild fish catches cannot be expected. Improved utilization of present harvest is therefore an important requisite for maintaining the contribution of fisheries to valuable food supplies.

This document is primarily focused on the Hazard Analysis Critical Control Point (HACCP) system which is at present recognized as the best system for assuring safety and sensory quality of food products. Additionally, the HACCP system is aimed at reducing the failure costs in the fishery industry, including the reduction of post-harvest losses.

The HACCP system is at the basis of the new regulations on fish inspection adopted by the European Economic Community (EEC), USA, Canada and a number of developing countries. Such new regulations are very often characterized as "HACCP-based systems".

FAO attaches great importance to training, and since 1986 the Fish Utilization and Marketing Service through different projects, in particular the FAO/DANIDA Training Project on Fish Technology and Quality Control, has provided training on HACCP to more than 2 500 fish technologists in developing countries. As impressive as this may appear, a lot of work must still be done to cover the present needs of developing countries in this field. We hope that this publication will contribute to this need.

W. Krone
Assistant Director-General a.i.
(Fisheries Department)

TABLE OF CONTENTS

1. INTRODUCTION

Seafood has traditionally been a popular part of the diet in many parts of the world and in some countries constituted the main supply of animal protein. Today even more people are turning to fish as a healthy alternative to red meat. The low fat content of many fish species (white fleshed, demersal) and the effects on coronary heart disease of the n-3 polyunsaturated fatty acids found in fatty (pelagic) fish species are extremely important aspects for health-conscious people particularly in affluent countries, where cardiovascular disease mortality is high. However, consumption of fish and shellfish may also cause diseases due to infection or intoxication. Some of the diseases have been specifically associated with consumption of seafood while others have been of more general nature.

For the purpose of this paper, seafood includes both finfish, shellfish and cephalopods (octopus, squid). The term shellfish covers the bivalve molluscan shellfish (oysters, cockles, clams and mussels), the gastropods (periwinkles, sea-snails) and the crustacean shellfish (crab, lobster, shrimp).

Seafood differs from other types of food in a number of ways. Most seafood is still extracted from a 'wild' population, and the fishermen are hunters with no influence on handling of their prey **before** it is caught. Thus it is not possible to imitate the situation for slaughter animals, selecting only the most suitable specimens for slaughter and to rest and feed them well before killing. The seafood processor is limited in his choice of raw materials to what is available in respect of size, condition and fish species landed by the fishermen. It should also be emphasized that while the inner and outer surface of warmblooded animals (gastrointestinal tract, skin) represent specific ecological environments with a very specific microbiological flora, these environments are very different for fish and shellfish. The microbiological flora in the intestines of these cold blooded animals is quite different being psychrotrophic in nature and to some extent believed to be a reflection of general contamination in the aquatic environment. Furthermore, in filter feeding bivalve molluscan shellfish (i.e. oysters), an accumulation and concentration of bacteria and viruses from the environment is generally taking place. However, some seafood is processed in a modern fish industry which is a technologically advanced and complicated industry in line with any other food industry, and with the same risk of products being contaminated with pathogenic organisms or toxin.

The quality of our foods is of major concern to food processors and public health authorities. It has been estimated that there are more than 80 million cases per annum of food-borne illnesses in the USA (Miller and Kvenberg 1986) and that the cost of these illnesses is in the order of many billions of dollars per year (Todd 1989b). The economic losses due to spoilage are rarely quantified but a report by the US National Research Council Committee (FNB/NRC 1985) estimates that one-fourth of the world's food supply is lost through microbial activity alone. Thus, the need for control of quality of our food is well documented and, since the rate of food-borne illnesses is increasing, there is also an urgent need to improve the traditional or present means of assuring the quality of food.

The word "quality" embraces a lot of meanings such as safety, gastronomic delights, purity, nutrition, consistency, honesty (e.g. in labelling), value, product excellence. This book focuses mainly on safety aspects, but also sensory quality (spoilage) will be dealt with and included in the quality assurance programmes. Control options and prevention measures to be applied within the various types of processing will be discussed.

From this very outset, a distinction needs to be drawn between Quality Assurance and Quality Control. Unfortunately, these two terms have been used indiscriminately and the difference between them has become blurred. According to International Standards Organization (ISO 8402), **Quality Assurance (Q.A.)** are "all those planned and systematic actions necessary to provide adequate confidence that a product or service will satisfy given requirements for quality". In other words, Q.A. is a strategic management function which establishes policies, adapts programmes to meet established goals - and provides confidence that these measures are being effectively applied. **Quality Control (Q.C.)** on the other hand are "the operational techniques and activities that are used to fulfill requirements for quality" (ISO 8402), i.e. a tactical function which carries out the programmes established by the Q.A.

2. STATISTICS ON SEAFOOD-BORNE DISEASES

The true incidence of diseases transmitted by foods is not known. There are many reasons for this. In most countries there is no obligation to report on food borne diseases to public health authorities. In the few countries which have a reporting system there is severe underreporting. It has been estimated that as few as 1% of the actual cases of food borne diseases are recorded (Mossel 1982). This is because neither the victim nor the physician are aware of the etiological role of foods. Furthermore, the food responsible is often not available for analysis and the true vehicle for the disease agent is not identified. The statistics below are therefore only for identifying trends and areas of concern.

Between 1973 and 1987 a total of 7,458 foodborne disease outbreaks involving 237,545 cases were reported in the United States (Bean and Griffin 1990). A specific food vehicle was identified in only 3699 (50%) of the outbreaks. Of these food items seafood was the food most frequently associated with disease as shown in Table 2.1.

Table 2.1. Types of food associated with incidents[1)] of foodborne diseases.

Food	USA[2)] 1973-1987		Canada[3)] 1982-1983		Netherlands[4)] 1980-1981	
	No.	%	No.	%	No.	%
Seafood	753	10.1	148	7.6	60	8.7
Meat (beef and pork)	579	7.8	404	20.7	91	13.2
Poultry	253	3.4	194	9.9	18	2.7
Vegetables	241	3.3	138	7.1	15	2.2
Eggs	38	0.5	4	0.2	1	0.1
Bakery products	100	1.3	151	7.7	27	3.9
Dairy foods	158	2.1	157	8.1	36	5.2
Other	1577	21.1	496	25.4	435	63.3
Total known[5)]	3,699	49.6	1,692	86.7	683	99.5
Unknown	3,759	50.4	259	13.3	3	0.5
Grand total	7,458	100.0	1,951	100.0	686	100.0

1) An incident is an outbreak (2 or more persons become ill) or a single case involving one person.
2) Data from Bean and Griffin (1990)
3) Data from Todd (1989a)
4) Data from Beckers (1986)
5) Total of incidents for which vehicles were identified

In the 2-year period (1980-1981) 8.7% of all outbreaks in the Netherlands were seafood-borne (Beckers 1986). However, Turnbull and Gilbert (1982) pointed out that specific foods are frequently not identified in food-poisoning incidents, but where they have been, fish and shellfish are implicated in less than 3% of all the general and family outbreaks reported in Britain. The incidence rates reported above should be evaluated in the light of total food consumption. Thus, in the same period in the U.S.A., the meat consumption was approx. 10 times that of fish and consumption of poultry approx. 5 times that of fish (Valdimarsson 1989).

The etiological agents associated with the large number of foodborne disease outbreaks reported in the USA in the period 1973-1987 are shown in Table 2.2.

Table 2.2. Etiologic agents associated with 7,458 outbreaks (involving 237,545 cases) of foodborne diseases reported to Center for Disease Control, Atlanta, USA, 1973-1987. Data from Bean and Griffin (1990).

	Outbreak			Cases		
Disease agent	No.	% of total	% of known	No.	% of total	% of known
Bacterial pathogens	1,875	25	66	108,745	46	87
Virus	142	2	5	11,249	5	9
Parasites	142	2	5	1,250	<1	1
Biotoxins	511	7	18	2,500	1	2
Chemicals	171	2	6	1,250	<1	1
Unknown	4,617	62	-	112,551	47	-
Total	7,458	100	100	237,545	100	100

In the majority of outbreaks (62% of total) the disease agent was not identified. One reason for this could be lack of proper technique for identification of e.g. virus. When identification of the etiological agent has been successful, pathogenic bacteria are by far the most frequent disease agent identified.

Diseases associated with various types of seafood over the period 1970-84 have been analyzed by Bryan (1980, 1987). He found that "fish" was most frequently involved followed by bivalve molluscan shellfish and crustaceans. Unfortunately, the reports available do not include information on the type of seafood **products** which were vehicles for the disease outbreaks. Knowledge of the preservation principles involved (a_w, pH, smoke, preservatives etc.), packaging and preparation before eating (cooking) would have been very useful in evaluation of the hazards related to the various types of seafood.

A considerable number (18%) of disease outbreaks related to "fish"recorded in the U.S. were of unknown etiology (see Figure 2.1). Most common were intoxications related to biotoxins (ciguatera) and histamine, which accounted for two thirds of all recorded outbreaks. The rest (18%) were caused by various bacteria, parasites, virus and chemicals.

FISH

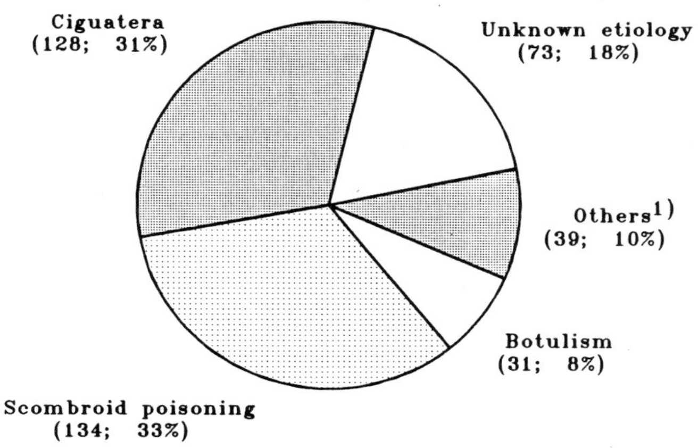

1) This group includes:
Staphylococcal intoxication
Shigellosis
Anisakiasis
C. perfringens gastroenteritis
Salmonellosis
Strept. pyogenes infection
Fish tape worm
Cholera
Typhoid fever
Pufferfish poisoning
V. parahaem. gastroenteritis
Hepatitis Non-B
Chemical poisoning

Figure 2.1. Diseases transmitted by fish in the United States from 1970 to 1984. (No. of outbreaks; %). Data from Bryan (1980) and Bryan (1987).

A total of 157 outbreaks of seafood borne diseases in the U.S. were related to consumption of molluscs. The great majority of them were of unknown etiology (see Figure 2.2). This fact should be viewed in the light of the great difficulties in diagnosing some of the viral diseases. Although only a few of the outbreaks recorded in Figure 2.2 are related to virus, there is no doubt that most illness associated with molluscs is mainly of viral origin.

Crustaceans were implicated as vehicle in a total of 63 outbreaks in U.S. during 1970-1984. More than one third of the outbreaks were of unknown etiology, but when the disease agent was identified, it was always a pathogenic bacterium (see Figure 2.3).

In a later study, Bean and Griffin (1990) analyzed the etiologic agents and food vehicles associated with 7,458 outbreaks (involving 237,545 cases) of foodborne diseases reported to Center for Disease Control, U.S.A. between 1973 and 1987. The disease agent was identified in only 2,841 outbreaks as shown in Table 2.2.

MOLLUSCS

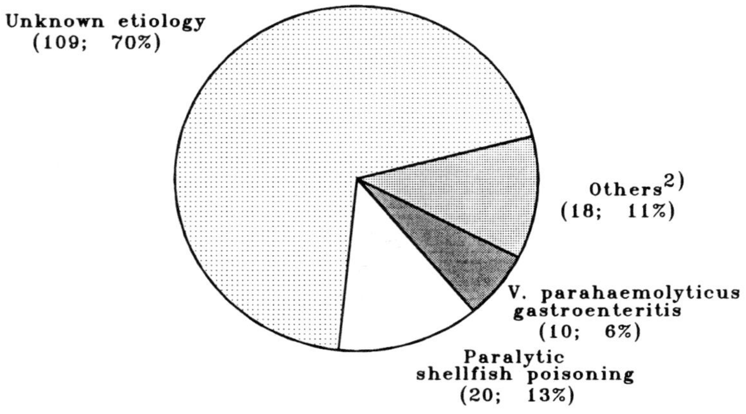

2) This group includes:
Staphylococcal intoxication
Shigellosis
Hepatitis A
Salmonellosis
Streptococcal infection
E. coli gastroenteritis
Cholera
A. hydrophila
B. cereus gastroenteritis
Hepatitis Non-B

Figure 2.2. Diseases transmitted by molluscs in the United States from 1970 to 1984. (No. of outbreaks; %). Data from Bryan (1980) and Bryan (1987).

CRUSTACEANS

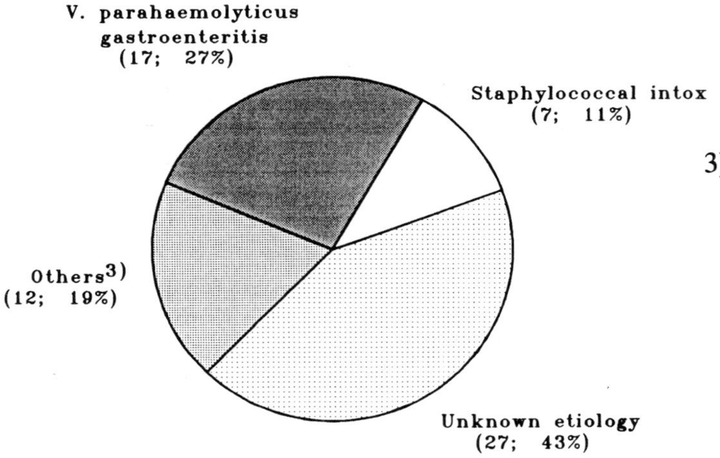

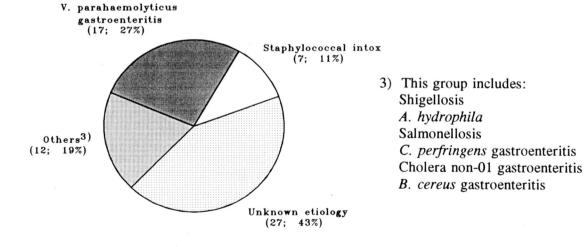

3) This group includes:
Shigellosis
A. hydrophila
Salmonellosis
C. perfringens gastroenteritis
Cholera non-01 gastroenteritis
B. cereus gastroenteritis

Figure 2.3. Diseases transmitted by crustaceans in the United States from 1970 to 1987. (No. of outbreaks; %). Data from Bryan (1980) and Bryan (1987).

Table 2.3. Etiologic agents associated with finfish (540 outbreaks) and shellfish (213 outbreaks) as vehicle in outbreaks of seafood borne diseases in U.S.A. in the period 1973-1987. Data from Bean and Griffin (1990).

Disease agent	Outbreaks	
	Finfish (%)	Shellfish (%)
Bacterial pathogens	10.0	17.0
Virus	0.2	5.2
Parasites	1.0	0.0
Biotoxin	80.0	9.8
Chemicals	0.7	0.5
Unknown	8.1	67.5

The data in Table 2.3 confirm that seafood borne diseases transmitted by finfish first of all are related to biotoxins and bacterial pathogens, while in the majority of shellfish transmitted diseases the disease agent has not been identified, but are probably viral.

3. QUALITY ASPECTS ASSOCIATED WITH SEAFOOD

In this chapter only the quality aspects related to safety and spoilage of seafood is discussed. The various disease agents which have been associated with consumption of seafood are listed and a few characteristics relevant to the evaluation of the hazards and risks related to their presence on fish and fish products are presented. The processes leading to spoilage and control options for disease agents as well as spoilage processes are briefly outlined.

3.1. PATHOGENIC BACTERIA

Seafood borne pathogenic bacteria may conveniently be divided into two groups as shown in Table 3.1.

Table 3.1. Seafoodborne pathogenic bacteria

		Mode of action		Heat stability of toxin	Minimum infective dose
		Infection	Pre-formed toxin		
Indigenous bacteria (Group 1)	*Clostridium botulinum*		+	low	-
	Vibrio sp.	+			high
	V. cholerae				-
	V. parahaemolyticus				($>10^6$/g)
	other vibrios[1]				-
	Aeromonas hydrophila	+			Not known
	Plesiomonas shigelloides	+			Not known
	Listeria monocytogenes	+			Not known / variable
Non-indigenous bacteria (Group 2)	*Salmonella* sp.	+			from $<10^2$ to $>10^6$
	Shigella	+			10^1- 10^2
	E. coli	+			10^1- 10^3 [2]
	Staphylococcus aureus		+	high	-

1) Other vibrios are: *V. vulnificus, V. hollisae, V. furnsii, V. mimicus, V. fluvialis.*
2) For verotoxin-producing strain 0157:H7

3.1.1. Indigenous bacteria (Group 1)

The bacteria belonging to the group 1 are common and widely distributed in the aquatic environments in various parts of the world. The water temperature is naturally having a selective effect. Thus the more psychrotrophic organisms (*C. botulinum* and *Listeria*) are common in Arctic and colder climates, while the more mesophilic types (*V. cholerae, V. parahaemolyticus*) are representing part of the natural flora on fish from coastal and estuarine environments of temperate or warm tropical zones.

It should be emphasized, however, that all the genera of pathogenic bacteria mentioned above contain non-pathogenic environmental strains. For some organisms it is possible to correlate between certain characteristics and pathogenicity (e.g. the Kanagawa-test for *V. parahaemolyticus)* while in others (e.g. *Aeromonas* sp.) there are no known methods available.

While it is true that all fish and fish products which have not been subject to bactericidal processing, may be contaminated with one or more of these pathogens, the level of contamination is normally quite low, and it is unlikely that the numbers naturally present in uncooked seafood are sufficient to cause disease. An exception is the cases when pathogens are concentrated due to filtration (molluscs). On the other hand high levels of group 1 bacteria may be found on fish products as a result of growth. This situation is constituting a serious hazard with a high risk of causing illness. Growth (and possible toxin production) must therefore be prevented. Some of the growth requirements of group 1 organisms are listed in Table 3.2. Some essential characteristics concerning each of the listed organisms are discussed below.

Clostridium botulinum

C. botulinum is widely distributed in soil, aquatic sediments and fish (Huss 1980, Huss and Pedersen 1979) as shown in Figure 3.1.

Human botulism is a serious but relatively rare disease. The disease is an intoxication caused by a toxin pre-formed in the food. Symptoms may include nausea and vomiting followed by a number of neurological signs and symptoms: visual impairment (blurred or double vision), loss of normal mouth and throat functions, weakness or total paralysis, respiratory failure, which is usually the cause of death.

Epidemiology and risk assessment

Examination of 165 outbreaks of botulism caused by fish products showed that the lightly preserved products (smoked, fermented) represented by far the most dangerous group as shown in Table 3.3.

Table 3.2. Growth limiting factors and heat resistance of pathogenic bacteria normally occurring on seafood (Group 1 - Indigenous bacteria). Data are adapted from Doyle (1989), Buckle (1989), Farber (1986) and Varnam and Evans (1991).

Pathogenic bacteria	Temperature (°C)		pH	a_w	NaCl (%)	Heat resistance
	minimum	optimum	minimum	minimum	maximum	
C. botulinum						
proteolytic type A,B,F	10	ca. 35	4.0 - 4.6	0.94	10	D_{121} of spores = 0.1-0.25 min.
non-proteolytic type B,E,F	3.3	ca. 30	5.0	0.97	3-5	$D_{82.2}$ = 0.15-2.0 min. in broth D_{80} = 4.5-10.5 min. in products with high protein and fat content[6]
Vibrio sp.	5-8	37	5.0			D_{71} = 0.3 min.[1]
V. cholerae	5	37	6.0	0.97	<8	D_{55} = 0.24 min.[2]
V. parahaemolyticus	5	37	4.8	0.93	8 - 10	60°C for 5 min. gave 7 $\log_{10}$ decline for *V. parahaemolyticus*
V. vulnificus	8	37	5.0	0.94	5	
Aeromonas sp.	0 - 4	20-35	4.0		4 - 5	D_{55} = 0.17 min. [5]
Plesiomonas sp.	8	37	4.0		4 - 5	60°C/30 min. no survival[7]
Listeria monocytogenes	1	30-37	5.0	0.92[4]	10	D_{60} = 2.4 16.7 min. in meat products[3] D_{60} = 1.95 - 4.48 min. in fish (Figure 3.3).

1) Shultz et al. (1984).
2) Delmore and Crisley (1979).
3) Farber and Peterkin (1991).
4) Nolan et al. (1992)
5) Condon et al. (1992)
6) Conner et al. (1989)
7) Miller and Koburger (1986)

Table 3.3. Type of fish products causing botulism. Data in this Table are taken from Huss (1981) and are representing outbreaks of botulism in Canada, Japan, USA, USSR, and Scandinavia over approx. 25 years (period 1950-1980).

Fish product	Process used	No. of outbreaks
Lightly preserved	smoked	10
	fermented	113
Semi-preserved	salted	9
	pickled	8
Fully preserved	canned	5
Unknown		20
Total		165

In contrast it should be noted that fresh and frozen fish has never been shown to cause human botulism. This is probably due to the fact that fresh fish normally spoil before becoming toxic. The ultimate safeguard is very low heat stability of botulinum toxin (Huss 1981, Hauschild 1989), which means that normal household cooking will destroy any preformed toxin. Thus the risk is clearly associated with foods that do not require cooking immediately before consumption.

Disease control

Botulism may be prevented by inactivation of the bacterial spores in heat-sterilized, canned products or by inhibiting growth in all other products. *C. botulinum* is classified into toxin types from A-G, and the types pathogenic to man can conveniently be divided into 2 groups:

1. The proteolytic types A and B, which are also heat-resistant, mesophilic and NaCl-tolerant.

2. The non-proteolytic types E, B and F, which are heat-sensitive, psychrotrophic and NaCl-sensitive. It is primarily the non-proteolytic types which are found on fish and fish products.

Canning processes have generally been designed to destroy a large number of the heat-resistant *C. botulinum* types. Thus the "botulinum cook" has been defined as equivalent to 3 min. at 121°C. This value is also called the F_0 value or the "process value". The F_0 value required for canned fish product is equivalent to 12-decimal reductions of *Clostridium botulinum* spores. Using the highest known D-values (0.25 min at 121°C) the F_0 is therefore equal to $12 \times 0.25 = 3$. This is the so-called 12 D-concept designed to reduce the bacterial load of one billion spores in each of 1000 cans to one spore in a thousand cans.

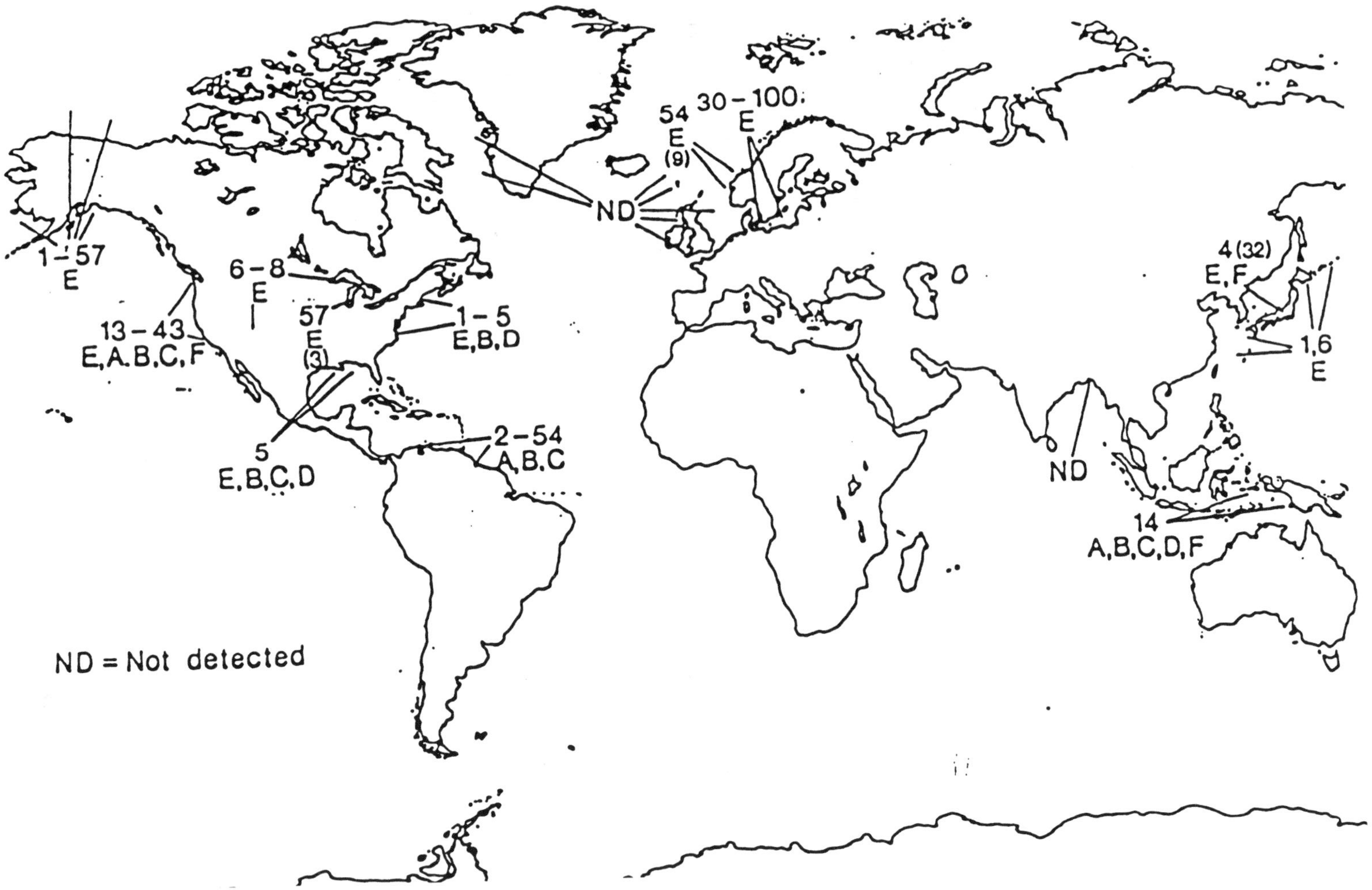

Figure 3.1. Incidence (%) of *C. botulinum* in fish.
Letters A-F indicate the presence of *C. botulinum* type A-F.
For references to the various surveys see Huss (1980).

In contrast, the D-value for the non-proteolytic group is much lower. Based on data presented by Angelotti (1970) a wet heat process of 82,2°C for 30 min. should destroy approx. 10^7 spores. Commercial heat pasteurization (sous-vide products, hot smoking) may therefore not be sufficient to kill all spores and safety of these products must be based on full control of growth - and toxin production.

Some of the most important limitations for growth for *C. botulinum* are listed in Table 3.2. Although it is shown that the non-proteolytic strains can grow in up to 5% NaCl this is only the case under optimal conditions. In fish products stored at low temperature (10°C), 3% NaCl in the water phase is sufficient to inhibit growth of type E for at least 30 days (Cann and Taylor 1979). Table 3.4 summarize the most important safety aspects of the various types of fish products.

Vibrio sp.

Most vibrios are of marine origin and they require Na^+ for growth. The genus contains a number of species which are pathogenic to man as listed in Table 3.1. *V. cholerae* occurs in two serotypes, the 01 and the non-01, and the 01 serotype occurs in two biovars: the classic and the El tor. The classical biovar, serovar 01 is today restricted to parts of Asia (Bangladesh), and most cholera is caused by the El tor biovar. The pathogenic species are mostly mesophilic, i.e. generally occurring (ubiquitous) in tropical waters and in highest numbers in temperate waters during late summer or early fall.

The diseases associated with *Vibrio* sp. are characterized by gastro-enteritic symptoms varying from mild diarrhea to the classical cholera, with profuse watery diarrhea. One exception is infections with *V. vulnificus*, which are primarily characterized by septicaemias.

The mechanisms of pathogenicity for the vibrios is not entirely clear. Most vibrios produce powerful enterotoxins and as little as 5μg cholera toxin (CT) administered orally caused diarrhea in human volunteers (Varnam and Evans 1991). A number of other toxins are produced by *V. cholerae*, including a hemolysin, a toxin similar to tetrodotoxin and one similar to shiga-toxin. Pathogenic strains of *V. parahaemolyticus* are known to produce a thermostable direct hemolysin (Vp-TDH), which are responsible for the Kanagawa-reaction, but it is now documented that also Kanagawa-negative *V. parahaemolyticus* are able to produce disease (Varman and Evans 1991).

The named pathogenic *Vibrio* sp. are not always pathogenic. The majority of environmental strains lack the necessary colonization factors for adherence and penetration, appropriate toxins or other virulence determinants necessary to cause disease.

In recent years it has been demonstrated that vibrios are able to respond to adverse environmental conditions by entering a viable, but non-culturable phase (Colwell 1986). When the bacteria are exposed to adverse conditions of salinity, temperature or nutrient deprivation, they can be reversibly injured and they cannot be detected by standard bacteriological methods. However, when given optimal conditions they can return to normal "culturable" state.

Table 3.4. Botulinogenic properties of fish products (after Huss 1981).

Fish Product	Factors adding to botulism hazard	Factors reducing botulism hazard	Safety of product based on:	Classification
Fresh and frozen	Vacuum packing	Traditional chill storage Putrefaction before toxin is produced	Cooking before being eaten	No risk
Pasteurized	Prolonged storage life Toxin produced before putrefaction Vacuum packing Poor hygiene	Chill storage (< 3°C) Synergistic aerobic flora eliminated	Cooking before being eaten Chill storage	No risk if cooked High risk if not cooked
Cold-smoked	Same as above Not cooked before being eaten No tradition for chill storage	Chill storage Salting (NaCl concentration >3%) High redox-potential in unspoiled products	Chill storage Process control (Raw material, salting when applicable)	High risk
Fermented	Fermentation may be slow High temperature during fermentation. Not cooked before being eaten	Salting (NaCl concentration >3% in brine) Chill storage Low pH	Process control Chill storage	High risk
Semi-preserved	Not cooked before being eaten	Application of salt, acid etc. Chill storage	Process control	Low risk
Fully preserved	Not cooked before being eaten Packed in closed cans	Autoclaving	Process control (Autoclaving, closing of cans)	Low risk

An obvious implication of this phenomenon is that routine examinations of environmental samples for these pathogens can be negative, while virulent bacteria are in fact present.

Epidemiology and risk assessment

Historically cholera is an illness of the poor and undernourished, but this is to some extend due to low standards of hygiene. In the case of cholera, water and fecal contamination of water is of major importance in spread of the disease, but food is becoming increasingly important.

A wide variety of foods has been involved in transmission of cholera, including soft drinks, fruits and vegetables, milk, locally brewed beer as well as millet gruel (Varnam and Evans 1991). However, raw, uncooked, or cross-contaminated cooked shellfish has been established as the major vehicle for *V. cholerae* 01 and non-01 (Morris and Black 1985). Outbreaks of *V. parahaemolyticus* has most often been associated with cross-contamination or time/-temperature abuse of cooked seafood. An exception is Japan, where raw finfish is the most common vehicle of infection with *V. parahaemolyticus*. For all other vibrios, consumption of raw shellfish, especially oysters, is the major cause of infection.

An important aspect is the remarkable growth rate exhibited by vibrios in raw fish, even at reduced temperatures. This allows relatively low initial numbers to increase dramatically under improper conditions of harvesting, processing, distribution and storage.

Disease control

Inadequate sanitation and lack of safe water are major causes of cholera epidemics. Therefore cholera can only be reliably prevented by ensuring that all populations have access to adequate excreta disposal systems and safe drinking water. Following the recent outbreak of cholera in South and Central America, WHO (1992) has issued the following recommendations on water supply and sanitation for prevention and control of cholera:

Water supply - WHO recommendations:

1. Drinking water should be adequately disinfected; procedures for disinfection in distribution systems and rural water systems should be improved.

2. Tablets releasing chlorine or iodine may be distributed to the population with instructions on their use.

3. Where chemical treatment of water is not possible, health educators should stress that water for drinking (as well as for washing of hands and utensils) should be boiled before use.

4. Water quality control should be strengthened by intensifying the surveillance and control of residual chlorine, and the conduct and analysis of bacteriological tests, in different points in production and distribution systems.

Sanitation - WHO recommendations:

1. Quality control in sewage treatment plants should be strengthened.

2. The use of treated waste water for irrigation should be carefully controlled, following national and international guidelines.

3. Large-scale chemical treatment of waste water is very rarely justified, even in emergencies, because of the high cost, uncertain effect, and possible adverse impact on the environment and health.

4. Health education should emphasize the safe disposal of human faeces:

   - All family members should use a latrine or toilet that is regularly cleaned and disinfected; and
   - Faeces of infants and children should be disposed of rapidly in a latrine or toilet, or by burying them.

Vibrios are easily destroyed by heat. Thus proper cooking is sufficient to eliminate most vibrios. However Blake et al. (1980) found *V. cholerae* 01 to survive boiling for up to 8 min. and steaming for up to 25 min. in naturally contaminated crabs. Thus the commercial practice of heat-shocking oysters in boiling water to facilitate opening is not enough to ensure safety.

At suitable temperatures growth of vibrios can be very rapid. Generation times as short as 8-9 min. have been observed under optimal conditions (37°C). At lower temperatures growth rates are reduced, but it was reported by Bradshaw et al. (1984) that initial concentrations of 10^2 cfu/g *V. parahaemolyticus* in homogenized shrimp increased to 10^8 cfu/g after 24 h at 25°C. These results demonstrate that proper refrigeration is essential in controlling such extravagant growth.

Low temperature storage has been proposed as a means of eliminating pathogenic vibrios from food. However, this method is not of sufficient reliability for commercial application. Mitscherlich and Marth (1984) have reported the survival times of *V. cholerae* shown in Table 3.5.

Table 3.5. Survival of *V. cholerae*. Data from Mitscherlich and Marth (1984).

Food	Survival times (days)
Fish stored at 3-8°C	14-25
Ice stored at -20°C	8
Shrimp, frozen	180
Vegetables in a moist chamber, 20°C	10
Carrots	10
Cauliflower	20
River water	210

Aeromonas sp.

The genus *Aeromonas* has been classified with the family *Vibrionaceae* and contains species pathogenic to animals (fish) and man. In recent years the motile *Aeromonas* sp., particularly *A. hydrophila* has received increasing attention as a possible agent of foodborne diarrheal disease. However, the role of *Aeromonas* as an enteric pathogen is not fully clarified.

Aeromonas is ubiquitous in freshwater environments, but may also be isolated from saline and estuarine waters (Knøchel 1989). This organism may also be readily isolated from meat, fish and seafood, ice-cream and many other foods as reviewed by Knøchel (1989). Indeed the organism has been identified as the main spoilage organism of raw meat (Dainty et al. 1983), raw salmon (Gibson 1992) packed in vacuum or modified atmospheres, and fish from warm, tropical waters (Gram et al. 1990, Gorczyca and Pek Poh Len 1985).

Species of *Aeromonas* produce a wide range of toxins such as cytotoxic enterotoxin, hemolysins and a tetrodotoxin-like sodium channel inhibitor (Varnam and Evans 1991). However, the role of these toxins in producing disease in man is unresolved and currently no method is available for differentiating between apathogenic environmental strains and pathogenic strains. Thus there is no evidence that toxins preformed in food play any role, and the association between eating fish and shellfish and *Aeromonas*-infection is at best circumstantial (Ahmed 1991).

Some growth limiting factors for *Aeromonas* are shown in Table 3.2. While minimum growth temperature for clinical strains is about +4°C (Palumbo et al. 1985) environmental strains and food isolates have been shown to grow at 0°C (Walker and Stringer 1987). *Aeromonas* is very sensitive to acid conditions and to salt and growth is unlikely to be a problem in foods where pH is less than 6.5 and the NaCl content greater than 3.0%.

Plesiomonas sp.

Also the genus *Plesiomonas* is placed in the family of Vibrionaceae. Like other members of this family is *Plesiomonas* widespread in nature but mostly associated with water, both fresh water and seawater (Arai et al. 1980). Due to its mesophilic nature (see Table 3.2) there is a marked seasonal variation in the numbers isolated from waters being much higher during warmer periods. Transmission by animals and intestines of fish is common, and it is likely that fish and shellfish is the primary reservoir of *Plesiomonas shigelloides* (Koburger 1989).

Plesiomonas sp. may cause gastroenteritis with symptoms varying from mild illness of short duration to severe diarrhea (shigella-like or cholera-like). However, it is possible that only a few strains carry virulent characteristics since volunteers ingesting the organism do not always become ill (Herrington et al. 1987). As is the case for *Aeromonas*, there is currently no way to differentiate pathogenic from non-pathogenic *Plesiomonas* sp.

Growth limiting factors are presented in Table 3.2.

Listeria sp.

Six species of *Listeria* are currently recognized, but only three sp. *L. monocytogenes, L. ivanovii*, and *L. seeligeri* are associated with disease in humans and/or animals. However, human cases involving *L. ivanovii* and *L. seeligeri* are extremely rare with only four reported cases. *L. monocytogenes* is subdivided into 13 serovars on the basis of somatic (O) and flagellar (H) antigens. This subdivision is of limited value in epidemiological studies since most of the isolates belong to three serotypes. More valuable methods are phage-typing, isoenzyme-typing, and DNA-fingerprinting. The latter one has shown promising results (Facinelli et al. 1988, Bille et al. 1992, Gerner-Smidt and Nørrung 1992).

L. monocytogenes is widespread in nature. It can be isolated from soil, vegetation, foods including fish and fish products and domestic kitchens as reviewed by Lovett (1989), Ryser and Marth (1991) and Fuchs and Reilly (1992). Most of these environmental strains are probably non-pathogenic.

Other *Listeria* sp. than *L. monocytogenes* appear to be more common in tropical areas (Fuchs and Reilly 1992, Karunasagar et al. 1992).

Listeriosis is an infection with the intestines as the point of entry, but the infective dose is not known. Incubation period may vary from one day to several weeks. Virulent strains are capable of multiplying in the macrophages and produce septicemia followed by infection of other organs such as the central nervous system, the heart, the eyes and may invade the fetus of pregnant women. In healthy adults, listeriosis usually never develops beyond the primary enteric phase, which may be symptom-free or having only mild "flu-like" symptoms. Listeriosis is of particular risk and can be lethal for foetuses, pregnant women, neonates and immuno-compromised persons.

Epidemiology and risk assessment

Dairy products (milk, cheese, ice-cream, cream butter) have all been implicated in outbreaks of listeriosis. Also salads and vegetables have been involved. Contaminated food is increasingly recognized as an important vehicle of *L. monocytogenes*. Frequent isolations from seafood (Weagant et al. 1989, Rørvik and Yndestad 1991) and the demonstration of growth potential in chilled (+4°C) smoked salmon (Ben Embarek and Huss 1992, Guyer and Jemmi 1991, Rørvik et al. 1991, Fuchs and Reilly 1992) are evidence that seafood may be important in the transmission of *Listeria monocytogenes*. However, so far there has only been two documented cases of seafood involvement (Facinelli et al. 1989, Frederiksen 1991), and two cases where seafood involvement is suspected (Lennon et al. 1984, Riedo et al. 1990).

Disease control

Currently FDA in the US requires that *L. monocytogenes* be absent in ready-to-eat seafood products such as crab-meat or smoked fish. This restriction does not apply to raw products that will be cooked before eating (Ahmed 1991). Other countries have similar regulations, which are completely unrealistic, since e.g. cold smoked fish has not been subject to listericidal processing. Due to the ubiquitous nature of *L. monocytogenes* such products can not be guaranteed free of *L. monocytogenes*. The FDA is now considering possible changes in their policy (Archer 1992). Products will be classified according to known, established risks. A zero-tolerance will still be maintained for products which have received a listericidal treatment, as well as for products which have been directly implicated in a foodborne outbreak. Low number of *L. monocytogenes* may then be allowed in other types of products, particularly those in which the organism can be shown to die-off.

There is a general agreement between microbiologists that the presence in our food of low numbers of *L. monocytogenes* may need to be tolerated. However, Notermans et al. (1992) suggest that a limit of 100 *L. monocytogenes*/g is reasonable, while Skovgaard (1992) feel that >10 *L. monocytogenes*/g is likely to constitute a risk to man - particularly predisposed persons (very old, very young or immuno-suppressed). These quoted figures should be compared to the background level of *L. monocytogenes* in foods, which is approx. 1-10 *L. monocytogenes*/g (Skovgaard 1992). This means that little or no growth of *L. monocytogenes* in foods should be tolerated.

However, the quantitative level of L.monocytogenes contamination on fish products can be maintained at a very low level (< 1-10/g) by proper GMP and factory hygiene. *L. monocytogenes* is sensitive to sanitizing agents as reviewed by Ryser and Marth (1991). Thus chlorine-based, iodine-based, acid anionic, and quaternary ammonium-type sanitizers are effective against *L. monocytogenes* at concentrations of 100 ppm, 25-45 ppm, 200 ppm and 100-200 ppm, respectively.

Further disease control with products which have not been subject to listericidal processing rest with control of growth in the products. Some growth limiting factors are listed in Table 3.2. It will be noted that *L. monocytogenes* is difficult to control in chilled fish products such as, e.g., cold smoked fish. The organism can grow at temperatures down to +1°C, and it is tolerant to NaCl (up to 10% at neutral pH and 25°C). Nitrite is not inhibitory to *L. monocytogenes* at permitted levels unless there is an interaction with other inhibiting agents (Shahamat et al. 1980). Thus it was demonstrated by Ben Embarek and Huss (1993) that no growth of *L. monocytogenes* occurred in vacuum packed cold smoked salmon having 5,4% NaCl in water phase and stored at 5°C for 25 days. However, growth was demonstrated in non vacuum packed (Guyer and Jemmi 1991) and vacuum packed (Rørvik et al. 1991) cold smoked salmon (2.5 - 3.2% NaCl in water phase) and stored at 4°C. Differences in NaCl content and the strains used may explain why different results were seen in these experiments.

Listericidal processing consist primarily of heat treatment. The heat resistance of *L. monocytogenes* has been the subject of extensive investigation particularly for milk and dairy products as reviewed by Mackey and Bratchell (1989). The thermal death time curve (T DTC) for *L. monocytogenes* in cod and salmon was studied by Ben Embarek and Huss (1993). The results show a significantly higher heat resistance of *L. monocytogenes* in salmon fillets compared to cod fillets with D_{60} being 4.5 min. in salmon and 1.8 min. in cod. The z-values were in both cases ca. 6°C as shown in Figure 3.2 which is very similar to the z-value calculated by Mackey and Bratchell (1989).

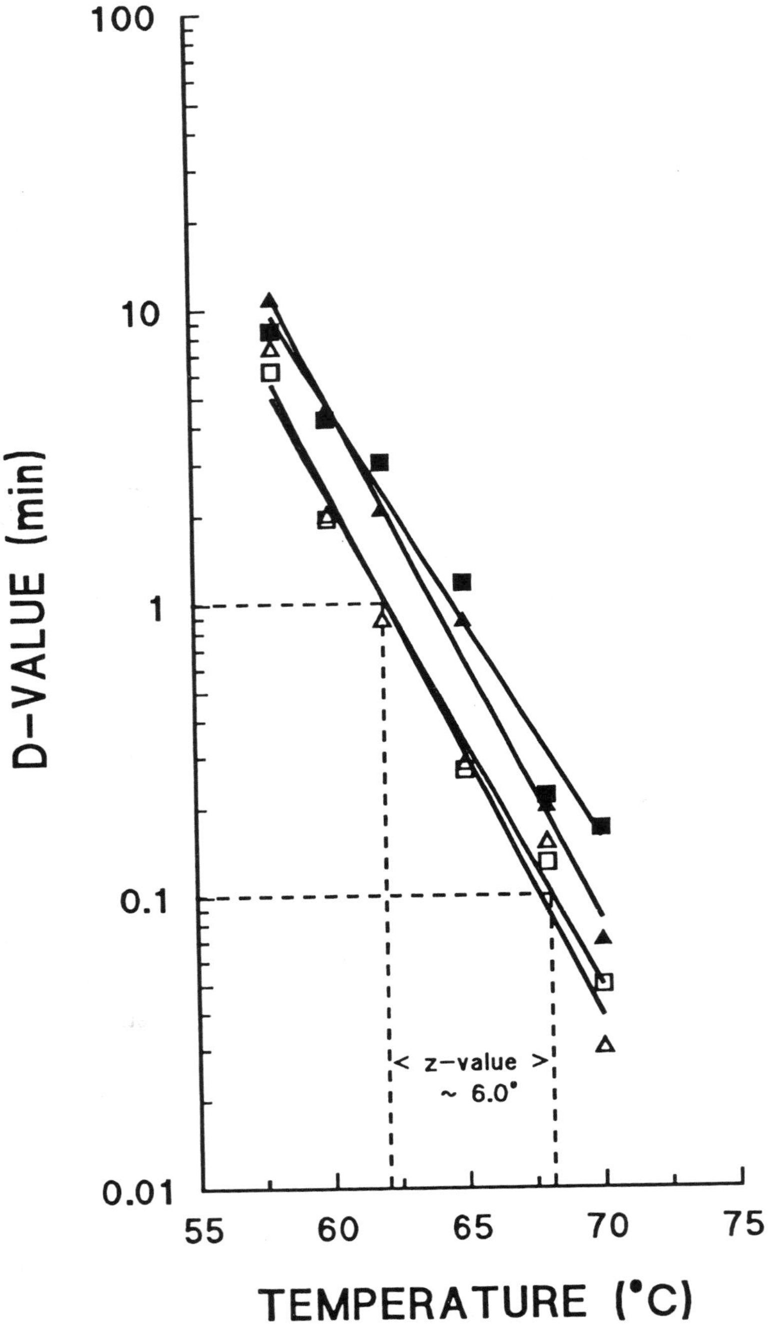

Figure 3.2. Heat resistance of *L. monocytogenes* in cod (open symbols) and salmon fillets (closed symbols). The test organisms were isolated from smoked salmon (squares) and clinical case of listeriosis (triangles) (Ben Embarek and Huss 1993).

3.1.2. Non-indigenous bacteria (Group 2)

Some of the growth requirements of group 2 organisms are listed in Table 3.6.

Salmonella sp.

Salmonella are members of the family *Enterobacteriaceae* and occur in more than 2000 serovars. These mesophilic organisms are distributed geographically all over the world, but principally occurring in the gut of man and animals and in environments polluted with human or animal excreta. Survival in water depends on many parameters such as biological (interaction with other bacteria) and physical factors (temperature). It has been demonstrated by Rhodes and Kator (1988) that both *E. coli* and *Salmonella* sp. can multiply and survive in the estuarine environment for weeks, while Jiménez et al. (1989) presented similar results on survival in tropical freshwater environments.

The principal symptoms of salmonellosis (non-typhoid infections) are non-bloody diarrhea, abdominal pain, fever, nausea, vomiting which generally appear 12-36 hours after ingestion. However, symptoms may vary considerably from grave typhoid like illness to asymptomatic infection. The disease may also proceed to more serious complications. The infective dose in healthy people varies according to serovars, foods involved and susceptibility of the individuals. There is evidence for a minimum infection dose (M.I.D.) of as little as 20 cells (Varnam and Evans 1991), while other studies have consistently indicated $>10^6$ cells.

Epidemiology and risk assessment

Salmonella is occurring commonly in domestic animals and birds and many are asymptomatic *Salmonella*-excreters. Raw meat and poultry are therefore often contaminated with this organism. Numerous surveys have been conducted as reviewed by D'Aoust (1989), showing that incidence varies according to species, agricultural- and processing practices. Between 50-100% of all samples of chicken carcasses are positive in most industrialized countries with intensively reared poultry, but also in other meats the contamination may be near to 100%. Also the contamination of raw milk, eggs and egg products with *Salmonella* is a long standing well known problem.

Contamination of shellfish with *Salmonella* due to growth in polluted waters has been a problem in many parts of the world. In a recent review by Reilly et al. (1992), evidence is presented that farmed tropical shrimps frequently contain *Salmonella*. However, it has also demonstrated that *Salmonella* in aquaculture shrimp products originate from the environment rather than as a result of poor standards of hygiene, sanitation, and poultry manure as feed.

Most literature reports indicate that seafood is a much less common vehicle for *Salmonella* than other foods, and fish and shellfish are responsible for only a small proportion of total number of *Salmonella* cases reported in U.S. and elsewhere (Ahmed 1991). Most prawns and shrimps are cooked prior to consumption and these products therefore pose minimal health risks to the consumer except by cross contamination in kitchens.

Table 3.6. Growth limiting factors and heat resistance of bacteria originating from the animal/human reservoir (Group 2 - non-indigenous bacteria). Data adapted from Doyle (1989), Buckle (1989), Varnam and Evans (1991) and Farber (1986).

Pathogenic bacteria	Temperature °C			pH	NaCl (%)	a_w	Heat resistance
	minimum	optimum	maximum	minimum	maximum	minimum	
Salmonella	5	37	45-47	4.0	4-5	0.94	D_{60} = 0.2-6.5 min.
Shigella	7-10	37	44-46	5.5	4-5		60°C/5 min.
E. coli	5-7	37	44-48	4.4	6	0.95	D_{60} = 0.1 min. D_{55} = 5 min.
Staphylococcus aureus	7	37	48	4.0	10-15	0.83	D_{60} = 0.43 -7.9 min.
Staphylococcus aureus toxin production	15	40-45	46	ca. 5.0	10	0.86	High heat stability of toxin

This is borne out by epidemiological evidence presented by Ahmed (1991), reporting on 7 outbreaks of seafoodborne salmonellosis in USA in the period 1978-1987. Three of these outbreaks were due to contaminated shellfish including 2 outbreaks after consumption of raw oysters harvested from sewage-polluted waters.

Shigella sp.

The genus *Shigella* is also a member of the *Enterobacteriaceae* and consists of 4 distinct species. This genus is specific host-adapted to humans and higher primates, and its presence in the environment is associated with fecal contamination. *Shigella* strains have been reported to survive for up to 6 months in water (Wachsmuth and Morris 1989).

Shigella is the cause of shigellosis (earlier name was bacillary dysentery), which is an infection of the gut. Symptoms vary from asymptomatic infection or mild diarrhea to dysentery, characterized by bloody stools, mucus secretion, dehydration, high fever and severe abdominal cramps. The incubation period for shigellosis is 1-7 days and symptoms may persist for 10-14 days or longer. Death in adults is rare, but the disease in children can be severe. In tropical countries with low standards of nutrition, shigella diarrhea accounts for the death of at least 500.000 children every year (Guerrant 1985).

Epidemiology and risk assessment

The great majority of cases of shigellosis is caused by direct person-to-person transmission of the bacteria via the oral-faecal route. Also waterborne transmission is important, especially where hygiene standards are low.

However, food, including seafood (shrimp-cocktail, tuna salads) have also been the cause of a number of outbreaks of shigellosis. This has nearly always been as a result from contamination of raw or previously cooked foods during preparation by an infected, asymptomatic carrier with poor personal hygiene.

Escherichia coli

E. coli is the most common aerobic organism in the intestinal tract of man and warm blooded animals. Generally the *E. coli* strains that colonize the gastrointestinal tract are harmless commensals, or they play an important role in maintaining intestinal physiology. However, within the specie there are at least 4 types of pathogenic strains:

1.	enteropathogenic *E. coli*	(EPEC)
2.	enterotoxigenic *E. coli*	(ETEC)
3.	enteroinvasive *E. coli*	(EIEC), shiga-dysentery-like *E.coli*
4.	enterohaemorrhagic *E.coli* / verocytoxin producing *E.coli*	(EHEC) / (VTEC) or *E. coli* 0157:H7

Serotyping as well as phage typing and genetic methods are used in epidemiological studies to separate among the various *E. coli* types, but there are no specific phenotypic marker to separate between pathogenic and non pathogenic strains. However, some atypical properties such as being lactose-negative or failure to produce indole at 44°C are more common between the pathogenic strains (Varnam and Evans 1991). VTEC do not grow at all on selective media at 44°C.

Clearly *E. coli* may be isolated in environments polluted by faecal material or sewage, and the organism can multiply and survive for a long time in this environment (Rhodes and Kator 1988, Jiménez et al. 1989). However, recently it has been demonstrated that *E. coli* also can be found in unpolluted warm tropical waters, where it can survive indefinitely (Hazen 1988, Fujioka et al. 1988, Toranzos et al. 1988).

Pathogenic *E. coli* strains are producing diseases of the gut which may vary in severity from extremely mild to severe and possibly life-threatening depending on a number of factors such as type of pathogenic strains, susceptibility of victim and degree of exposure.

Epidemiology and risk assessment

There is no indication that seafood is an important source of *E. coli* infection (Ahmed 1991). Most infections appear to be related to contamination of water or handling of food under unhygienic conditions.

Control of *Enterobacteriaceae*

The *Enterobacteriaceae*, (*Salmonella, Shigella, E. coli*) are all occurring on fish products as a result of contamination from the animal/human reservoir. This contamination has normally been associated with fecal contamination or pollution of natural waters or water environments, where these organisms may survive for a long time (months) or through direct contamination of products during processing.

Good personal hygiene and health education of food handlers are therefore essential in the control of diseases caused by *Enterobacteriaceae*. Proper treatment (e.g. chlorination) of water and sanitary disposal of sewage are also essential parts in a control programme.

Risk of infection with *Enterobacteriaceae* can be minimized or eliminated by proper cooking before consumption. It is well established that the heat resistance of *Salmonella* is low, but also that it varies considerably with a_w and with the nature of solutes in the heating menstruum (D'Aoust 1989). Thus a markedly increased heat resistance has been recorded at low a_w. Examples of D-values in high a_w foods are quoted in Table 3.6 as well as other physical factors limiting the growth of *Enterobacteriaceae*. Thus the growth is generally inhibited in the presence of 4-5% NaCl. Increased inhibition is seen at low temperature or reduced pH. The limiting water activity (a_w) for *Salmonella* in broth cultures are found to be 0.94 (Marshall et al. 1971).

Growth limiting factors for *Shigella* and some pathogenic *E. coli* are of no importance due to the low infective dose required to produce disease.

Current levels of *Salmonella* in various foods and increasing trends in human infections and foodborne outbreaks (D'Aoust 1989) underline that bacteriological testing and stringent bacteriological standards (zero tolerance limits) of most foods are insufficient measures in the control of salmonellosis. Even the microbial quality of harvest water appears not to be a good predictor of *Salmonella* contamination, because oysters removed from closed and open beds had same level of contamination (4%) and no correlation was observed between the presence of *E. coli* and *Salmonella* (D'Aoust et al. 1980).

Staphylococcus aureus

The staphylococci are ubiquitous organisms and can be found in water, air, dust, milk, sewage, floors, surfaces, all articles that come into contact with man and survive very well in the environment. However, the main reservoir and habitat is the animal/human nose, throat and skin. The human carrier rate may be up to 60% of healthy individuals with an average of 25-30% of the population being positive for enterotoxin-producing strains (Ahmed 1991).

The disease caused by *S. aureus* is an intoxication. Common symptoms, which may appear within 2-4 hours of consumption of contaminated foods are nausea, vomiting and sometimes diarrhea. Symptoms usually persist for no more than 24 hours, but in severe cases, dehydration can lead to shock and collapse.

Epidemiology and risk assessment

Seafood may be contaminated with *Staphylococcus* via infected food handlers or from the environment. More often the contamination is from an individual with an infection on hands or with a cold or sore throat.

S. aureus is mesophilic with a minimum growth temperature of 10°C, but higher temperatures are required for toxin production (>15°C). In contrast to the *Enterobacteriaceae*, but in common with *L. monocytogenes, S. aureus* is halotolerant and able to grow at water activities as low as 0.86. Minimum pH for growth is 4.5. The above minimum requirements are related to growth in laboratory media, when other factors are optimal. This is not always the case in food, where several limiting factors may be acting in combination. It should also be emphasized that staphylococci are poor competitors and do not grow well in the presence of other microorganisms. Thus the presence of staphylococci in raw, naturally contaminated food is of little significance. In contrast rapid growth and toxin production can take place in precooked seafood (shrimp) if recontaminated with *S. aureus* and time/ temperature conditions allow for growth.

S. aureus produces a number of enterotoxins, when growing in the food. These toxins are generally very resistant to proteolytic enzymes and heat. There have been no outbreaks reported from foods that have undergone normal canning procedures, but the heat applied in pasteurization and normal household cooking is not sufficient to destroy the toxin.

Disease control

Good sanitary conditions and temperature control is necessary to avoid contamination, growth and toxin production - particularly in precooked seafood.

3.2. VIRUS

The incidence of food-borne outbreaks of viral gastroenteritis is still unknown, but some authors believe that they are quite common. Progress has been slow in studying the viruses that infect the human gut and little is known about many of the important characteristics of enteric viruses. Cultivation of some virus (e.g. Hepatitis A virus, HAV) is now possible, but reliable methods for detection of viruses in food is not available. However, techniques based on molecular biology such as RNA/DNA-probes and PCR (polymerase chain reaction) routines are rapidly being developed.

Viral disease transmission to human via consumption of seafood has been known since the 1950's (Roos 1956), and human enteric viruses appear to be the major cause of shellfish-associated disease. Presently there are more than 100 known enteric viruses which are excreted in human faeces and find their way into domestic sewage. However, only a few have shown to cause seafood-associated illness according to Kilgen and Cole (1991).

These are:
- Hepatitis - type A (HAV)
- Norwalk virus (small, round structured)
- Snow Mountain Agent
- Calicivirus
- Astrovirus
- Non-A and Non-B.

Viruses are inert outside the living host-cell, but they survive. This means that they do not replicate in water or seafood irrespective of time, temperature or other physical conditions. Their presence on seafood is purely as a result of contamination either via infected food handlers or via polluted water. Shellfish which are filter-feeders tend to concentrate virus from the water in which they are growing. Large amounts of water are passing through active shellfish (up to 1,500 l/day/oyster according to Gerba and Goyal (1978)) which means that the concentration of virus in the shellfish is much higher than in the surrounding water.

Epidemiology and risk assessment

The infective dose of viruses is probably much smaller than that of bacteria for causing foodborne disease (Cliver 1988). The minimum infection dose of some enteric viruses for man is close to the minimum dose detectable in laboratory assay systems using cell cultures (Ward and Akin 1983).

The animal/human bodies are the sources of enteric viruses. The viruses are found in large quantities in the faeces of infected persons a few days to several weeks after ingestion/ infection depending on the virus. Direct or indirect faecal contamination is the most common source of contamination of food.

The list of food vehicles in outbreaks of viral diseases is dominated by bivalve molluscs. However, another important vehicle involves ready to eat food prepared by infected food handlers. The available data show that almost any food that comes into contact with human hands and does not subsequently receive a substantial heat treatment, may transmit these viruses.

With only few exceptions, all reported cases of seafood-associated viral infections have been from consumption of raw or improperly cooked molluscan shellfish (Kilgen and Cole 1991). However, there is clear evidence that HAV has been transmitted by unsanitary practices during processing, distribution or food handling (Ahmed 1991). These seafood associated illnesses are very common. Each year 20,000 to 30,000 cases are reported to the Center of Disease Control (CDC) in U.S. (Ahmed 1991), and one of the largest outbreaks of foodborne illness ever reported, is the outbreak of hepatitis involving 290,000 cases in China in 1988. The investigation revealed that the source and mode of transmission were the consumption of contaminated and inadequately cooked clams (Tang et al. 1991).

The survival of viruses in the environment and in food is dependent on a number of factors such as temperature, salinity, solar radiation, presence of organic solids as reviewed by Gerba (1988). Thus, enteric viruses are able to survive for several months in seawater at temperatures $< 10°C$, which is much longer than e.g. coliform bacteria (Melnick and Gerba 1980). Thus, there is little or no correlation between presence of virus and the usually applied indicator bacteria for faecal pollution. All enteric viruses are also resistant to acid pH, proteolytic enzymes and bile salts in the gut. Hepatitis type A virus, being one of the more heat stable viruses, has an inactivation time of 10 min at 60°C (Eyles 1989). Thus virus is able to survive some commonly used culinary preparations (steaming, frying). Enteric viruses are also resistant to some common disinfectants (e.g. phenolics, quaternary ammonium compounds, ethanol) while the halogens (e.g. chlorine, iodine) inactivate enteric viruses in water and on clean surfaces. Ozone is highly effective in clean water (quoted after Eyles 1989).

Disease control

Prevention of foodborne viral disease relies on measures to prevent direct or indirect fecal contamination of food that will not receive a virucidal treatment before consumption.

Bivalve shellfish are fit for human consumption if harvested in pollution free waters or alternatively rendered fit by depuration in clean seawater or by cooking. However, there are considerable problems in such a programme:

- Monitoring of harvesting-areas has been based on bacterial indicators of pollution, which are known to be unreliable predictors of viral contamination (Richards 1985, Cliver 1988).

- Depuration technology may be inadequate on some occasions for removal of virus from shellfish (Eyles 1986, Gerba 1988) and there is no practical test to indicate that shellfish have been depurated effectively.

Contamination by food handlers can be prevented by good personal hygiene and health education as mentioned for control with *Enterobacteriaceae*. Food handlers must not handle food while suffering from intestinal infections and for at least 48 h after symptoms have disappeared. In cases of doubt, disposable gloves should be worn in critical operations, as viruses are difficult to remove from hands by washing and are resistant to many skin disinfectants (Eyles 1989).

3.3. BIOTOXINS

Marine biotoxins are responsible for a substantial number of seafood borne diseases. The toxins which are known are shown in Table 3.7.

Table 3.7. Aquatic biotoxins

Toxin	Where/when produced	Animal(s)/organ involved
Tetrodotoxin	in fish *ante mortem*	pufferfish (*Tetraodontidae*) mostly ovaries, liver, intestines
Ciguatera	Marine algae	> 400 tropical/subtropical fish sp.
PSP-paralytic shellfish poison	" "	filter feeding shellfish, mostly digestive glands and gonads
	" "	filter feeding shellfish
DSP-diarrhetic " "	" "	" " "
NSP-neurotoxic " "	" "	" " " (blue mussels)
ASP-amnesic " "		

The toxins and the diseases they can provoke have been described and reviewed by Taylor (1988), Hall (1991), WHO (1984a, 1989) and Todd (1993), which should be consulted for detailed information. Some of the more important aspects are discussed below.

Tetrodotoxin

Unlike all other biotoxins accumulating in the live fish or shellfish tetrodotoxin is not produced by algae. The precise mechanism in production of this very potent toxin is not clear, but apparently quite commonly occurring symbiont bacteria are involved (Noguchi et al. 1987, Matsui et al. 1989).

Tetrodotoxin is mainly found in the liver, ovaries and intestines in various species of pufferfish, the most toxic being members of the family *Tetraodontidae*, but not all species in this family contain the toxin. The muscle tissue of the toxic fish is normally free of toxin, but there are exceptions. Pufferfish poisoning causes neurological symptoms 10-45 minutes after ingestion. Symptoms are tingling sensation in face and extremities, paralysis, respiratory symptoms and cardiovascular collapse. In fatal cases death takes place within 6 hours.

Ciguatera

Ciguatera poisoning results from the ingestion of fish that have become toxic by feeding on toxic dinoflagellates, which are microscopic marine planktonic algae. The principal source is the benthic dinoflagellate *Gambierdiscus toxicus*, which is living around coral reefs closely attached to macroalgae. Increased production of toxic dinoflagellates are seen when reefs are disturbed (hurricanes, blasting of reefs etc.). More than 400 species of fish, all found in tropical or warm waters, have been reported to have caused ciguatera as shown in Figure 3.3 (Halstead 1978). The toxin accumulates in fish that feed on the toxic algae or larger carnivores that prey on these herbivores. Toxin can be detected in gut, liver and muscle tissue by means of mouse-assay and chromatography. Some fish may be able to clear the toxin from their systems (Taylor 1988).

Although the reported incidence of ciguatera poisoning is low (Taylor 1988), it has been estimated that the world-wide incidence may be in the order of 50,000 cases/year (Ragelis 1984). The clinical picture varies but onset time is a few hours after ingestion of toxin. Gastrointestinal and neurological systems are affected (vomiting, diarrhea, tingling sensation, ataxia, weakness). Duration of illness may be 2-3 days but some may also persist for weeks or even years in severe cases. Death results from circulatory collapse. Halstead (1978) has reported a case-fatality rate of about 12%.

Paralytical shellfish poisoning (PSP)

Intoxication after consumption of shellfish is a syndrome that has been known for centuries, the most common being paralytic shellfish poisoning (PSP). PSP is caused by a group of toxins (saxitoxins and derivatives) produced by dinoflagellates of the genera *Alexandrium, Gymnodinium* and *Pyrodinium*.

Historically, PSP has been associated with the blooming of dinoflagellates ($>10^6$ cells/litre), which may cause a reddish or a yellowish discolouration of the water. However, water discolouration may be caused by proliferation of many types of planktonic species which are not always toxic and not all toxic algae blooms are coloured.

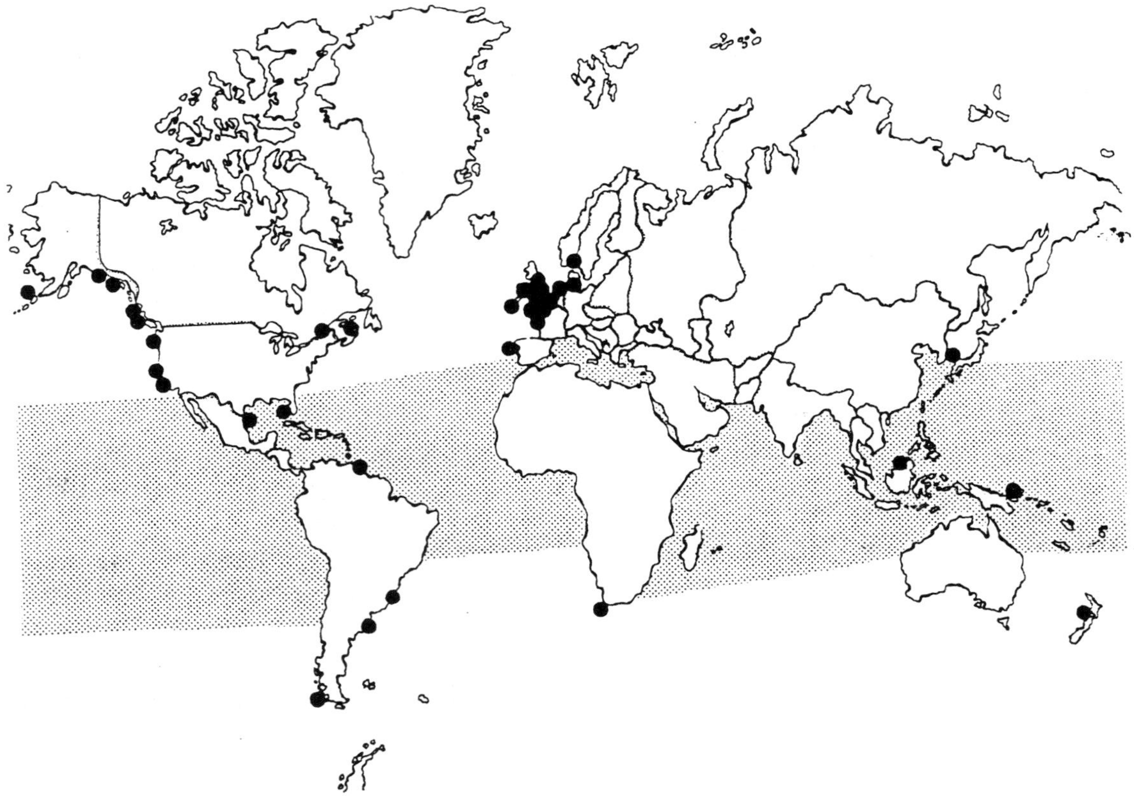

Figure 3.3. World distribution of outbreaks of paralytic shellfish poisoning (black spots) and ciguatera (shaded area). Data from WHO (1984a), Halstead and Schantz (1984) and Lupin (1992).

The dinoflagellates bloom as a function of water temperature, light, salinity, presence of nutrients and other environmental conditions. However, the precise nature of factors eliciting a toxic clone is unknown. Water temperature must be >5-8°C for blooms to occur. If temperatures decrease to below 4°C, the dinoflagellates will survive as cysts buried in the upper layers of the sediments. The worldwide occurrence of PSP is shown in Figure 3.3.

Mussels, clams, cockles and scallops that have fed on toxic dinoflagellates retain the toxin for varying periods of time depending on the shellfish. Some clear the toxin very quickly and are only toxic during the actual bloom, others retain the toxin for a long time, even years (Schantz 1984).

PSP is a neurological disorder, and the symptoms, include tingling, burning and numbness of lips and fingertips, ataxia, drowsiness, incoherent speech. In severe cases death occurs due to respiratory paralysis. Symptoms develop within 0.5-2 h of a meal and victims who survive more than 12 h generally recover.

Diarrhetic shellfish poisoning (DSP)

Thousands of cases of gastrointestinal disorders caused by diarrhetic shellfish poisoning (DSP) have been reported in Europe, Japan and Chile (WHO 1984a). The causative dinoflagellates which produce the toxins are within the genus *Dinophysis* and *Aurocentrum*. These dinoflagellates are widespread which means that this illness could also occur in other parts of the world. At least 7 toxins have been identified, including okadoic acid. Onset of disease is within half an hour to a few hours following consumption of shellfish which have been feeding on toxic algae. Symptoms are gastrointestinal disorder (diarrhea, vomiting, abdominal pain) and victims recover within 3-4 days. No fatalities have ever been observed.

Neurotoxic shellfish poisoning (NSP)

Neurotoxic shellfish poisoning (NSP) has been described in people who consumed bivalves that have been exposed to "red tides" of the dinoflagellate (*Ptychodiscus breve*). The disease has been limited to the Gulf of Mexico and areas off the coast of Florida. Brevetoxins are highly lethal to fish and red tides of this dinoflagellate is also associated with massive fish kills.

The symptoms of NSP resembles PSP except that paralysis does not occur. NSP is seldom fatal.

Amnesic shellfish poisoning (ASP)

Amnesic shellfish poisoning (ASP) has only recently been identified (Todd 1990, Addison and Stewart 1989). The intoxication is due to domoic acid, an amino acid produced by the diatom *Nitzschia pungens*. The first reported incidence of ASP occurred in the winter of 1987/88 in eastern Canada, where over 150 people were affected and 4 deaths occurred after consumption of cultured blue mussels.

The symptoms of ASP vary greatly from slight nausea and vomiting to loss of equilibrium and central neural deficits including confusion and memory loss. The short term memory loss seems to be permanent in surviving victims, thus the term amnesic shellfish poisoning.

Control of disease caused by biotoxins

The control of marine biotoxins is difficult and disease cannot be entirely prevented. The toxins are all of non-protein nature and extremely stable (Gill et al. 1985). Thus cooking, smoking, drying, salting does not destroy them, and one cannot tell from the appearance of fish or shellfish flesh whether it is toxic.

The major preventive measure is inspection and sampling from fishing areas and shellfish beds, and analysis for toxins. The mouse bioassay is often used for this purpose and confirmatory HPLC is done if death occurs after 15 min. If high levels of toxin are found, commercial harvesting is halted. It seems unlikely that it will ever be possible to control phytoplankton composition in growing areas, eliminating toxigenic species, and there is no reliable way to forecast, when a particular phytoplankton will grow and thus no way to predict blooming of toxigenic species (Hall 1991).

Removal of toxin by depuration techniques may have some potential, but the process is very slow and costly. There is also a risk that a small number of individuals decline to open and pump clean water through the system and therefore retain their original level of toxicity (Hall 1991).

To be effective, the monitoring requires reliable sampling plans and efficient means of detection of the toxins. Reliable chemical methods for detection of all toxins are at present available and should be developed. The sampling plan must take into consideration that toxicity of shellfish can increase from negligible to lethal levels in less than one week or even less than 24 h for blue mussels. Also the toxicity can vary within a growing location for shellfish according to geography, water currents and tidal activity.

The present situation regarding tolerances and methods of analysis to be used in a monitoring programme is shown in Table 3.8.

Table 3.8. Monitoring of biotoxins (WHO 1989)

Toxin	Tolerance	Method of analysis
Ciguatera	control not possible	No reliable method
PSP	80 μg/100g	Mouse bioassay, HPLC
DSP	0-60μg/100g	Mouse bioassay, HPLC
NSP	any detectable level/100 g is unsafe	Mouse bioassay. No chemical method
ASP	20 μg/g domoic acid	HPLC

3.4. BIOGENIC AMINES (HISTAMINE POISONING)

Histamine poisoning is a chemical intoxication following the ingestion of foods that contain high levels of histamine. Historically this poisoning was called scombroid fish poisoning because of the frequent association with scombroid fishes including tuna and mackerel.

Histamine poisoning is a world-wide problem occurring in countries where consumers ingest fish containing high levels of histamine. It is a mild disease; incubation period is very short (few minutes to few hours) and duration of illness is short (few hours). The most common symptoms are cutaneous such as facial flushing, urticaria, edema, but also the gastrointestinal tract may be affected (nausea, vomiting, diarrhea) as well as neurological involvement (headache, tingling, burning sensation in the mouth).

Histamine is formed in the fish *post mortem* by bacterial decarboxylation of the amino acid histidine as shown in Figure 3.4. The fish frequently involved are those with natural high content of histidine such as those belonging to the family *Scombridae* but also non-scombroid fish such as *Clupeidae* and mahi-mahi may be involved in histamine poisoning.

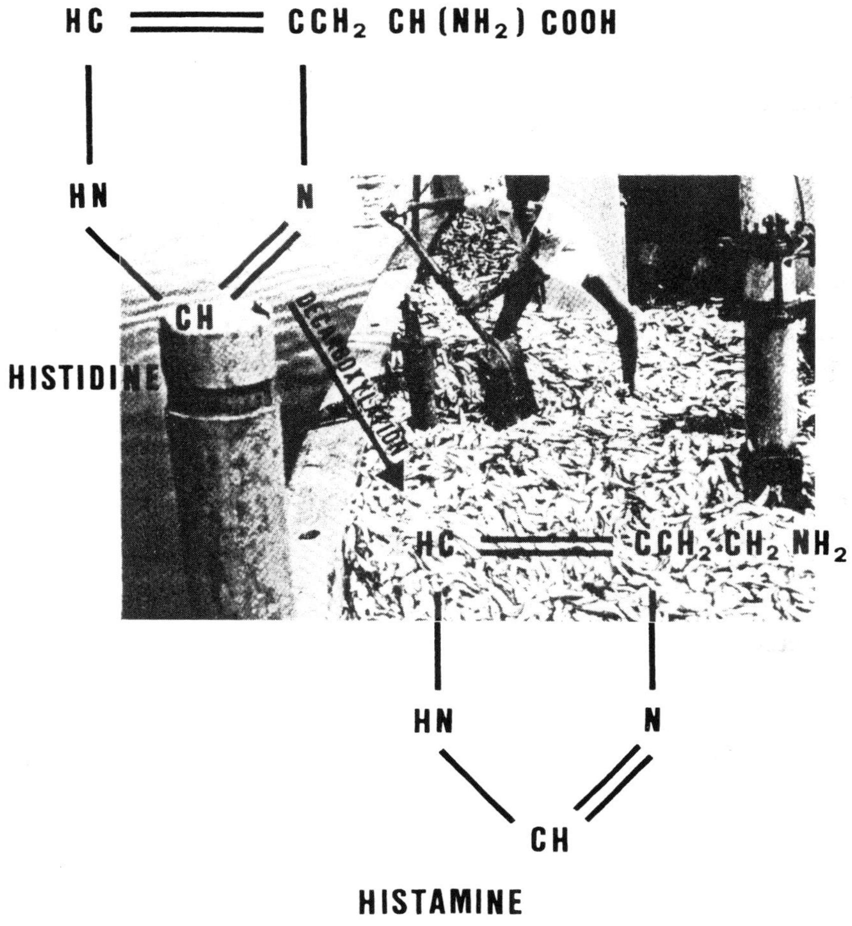

Figure 3.4. Chemical structure of histamine (photo: Pan and James 1985)

The histamine-producing bacteria are certain *Enterobacteriaceae*, some *Vibrio* sp. a few *Clostridium* and *Lactobacillus* sp. The most potent histamine producers are *Morganella morganii*, *Klebsiella pneumoniae* and *Hafnia alvei* (Stratten and Taylor 1991). These bacteria can be found on most fish, probably as a result of post-harvest contamination. They grow well at 10°C but at 5°C growth is greatly retarded and no histamine was produced by *M. morganii* when temperatures were <5°C at all times (Klausen and Huss 1987). However, large amounts of histamine were formed by *M. morganii* at low temperatures (0-5°C) following storage for up to 24 h at high temperatures (10-25°C) even though bacterial growth did not take place at 5°C and below.

Many studies agree that histamine producing bacteria are mesophilic. However, Ababouch et al. (1991) found considerable histamine production in sardines at temperatures <5°C, and van Spreekens (1987) has reported on histamine production by *Photobacterium* sp. which are also able to grow at temperatures <5°C.

The principal histamine producing bacteria *M. morganii* grow best at neutral pH, but they can grow in the pH range 4.7-8.1. The organism is not very resistant to NaCl, but at otherwise optimal conditions growth can take place in up to 5% NaCl. Thus histamine production by this organism is only a problem in very lightly salted fish products.

It should be emphasized that once the histamine has been produced in the fish, the risk of provoking disease is very high. Histamine is very resistant to heat, so even if the fish is cooked, canned or otherwise heat-treated before consumption, the histamine is not destroyed.

The evidence that histamine is causing disease is mostly circumstantial. High levels of histamine has consistently been found in samples implicated in outbreaks, and the symptoms noted in outbreaks are consistent with histamine as the causative agents. However, high intake of histamine does not always result in disease, even when "hazard action level" (50 mg/100 g for tunafish) is exceeded.

The human body will tolerate a certain amount of histamine without any reaction. The ingested histamine will be detoxified in the intestinal tract by at least 2 enzymes, the diamine oxidase (DAO) and histamine N-methyltransferase (HMT) (Taylor 1986). This protective mechanism can be eliminated if intake of histamine and/or other biogenic amines is very high, or if the enzymes are blocked by other compounds as shown in Figure 3.5.

Other biogenic amines such as cadaverine and putrescine which are known to occur in spoiled fish may therefore act as potentiators of histamine toxicity. Presumably inhibition of intestinal histamine catabolism will result in greater transport of histamine across cellular membranes and into the blood circulation.

Control of disease caused by biogenic amines

Low temperature storage and holding of fish at all times is the most effective preventive measure. All studies seem to agree that storage at 0°C or very near to 0°C limits histamine formation in fish to negligible levels.

Several countries have adapted regulations governing the maximum allowable levels of histamine in fish. Examples are shown in Table 3.9.

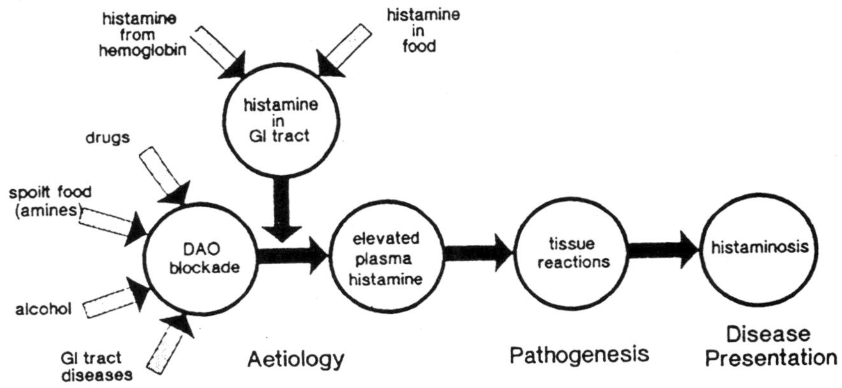

Figure 3.5. The disease concept of food-induced histaminosis (after Sattler and Lorenz 1990)

Table 3.9. Regulatory limits for histamine in fish.

	Hazard action level mg/100 g	Defect action level mg/100 g	Maximum allowable limit g/100g
USA (FDA)	50	10 - 20	-
EEC	-	10	20

3.5. PARASITES

The presence of parasites in fish is very common, but most of them are of little concern with regard to economics or public health. Reviews have been published by Healy and Juranek (1979), Higashi (1985) and Olson (1987).

However, more than 50 species of helminth parasites from fish and shellfish are known to cause disease in man. Most are rare and involve only slight to moderate injury but some pose serious potential health risk. The most important are listed in Table 3.10.

All the parasitic helminths have complicated life cycles. They do not spread directly from fish to fish but must pass through a number of intermediate hosts in their development. Very often sea-snails or crustaceans are involved as first intermediate host and marine fish as second intermediate host, while the sexually mature parasite is found in mammals as the final host.

In between these hosts, one or more free living stages may occur. Infection of human may be part of this life cycle or it may be a side track causing disruption of the life cycle as shown in Figure 3.6.

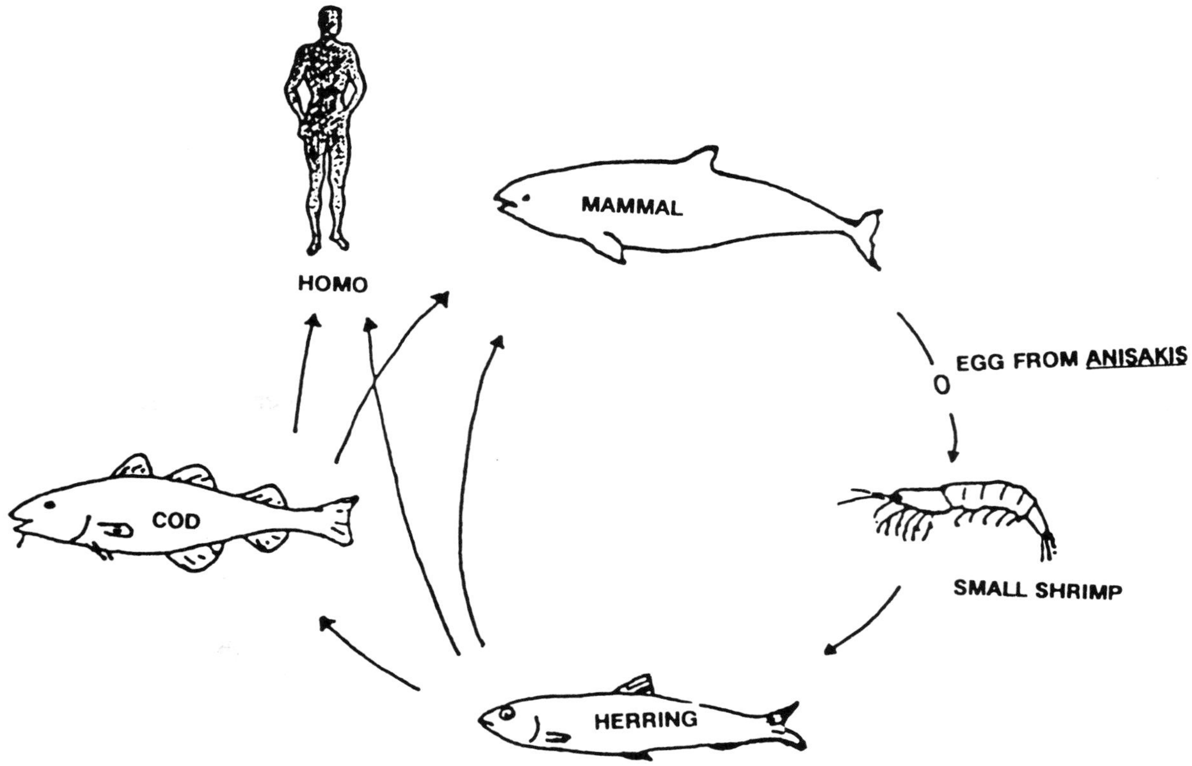

Figure 3.6. Life cycle of *Anisakis simplex*.

Nematodes

Round worms or nematodes are common and found in marine fish all over the world. The anisakis nematodes *A. simplex* and *P. dicipiens* commonly known as the herring worm and the cod worm have been intensively studied. They are typical round worms, 1-6 cm long, and if live worms are ingested by humans they may penetrate into the wall of the gastrointestinal tract and cause an acute inflammation ("herring worm disease"). The complete life cycle of *Anisakis* sp. is shown in Figure 3.6.

A number of other nematodes are found in freshwater fish. *Gnathostoma* sp. are the most important species found in Asia. The final hosts are cats and dogs but humans may be infected. Upon ingestion the larvae migrate from the stomach to various regions, most commonly to subcutaneous sites on the thorax, arms, head and neck, where the worms induce a creeping sensation and edema.

Table 3.10. Pathogenic parasites transmitted by fish and shellfish.

Parasite	Known geographical distribution	Fish and shellfish
Nematodes or round worms		
Anisakis simplex	North Atlantic	herring
Pseudoterranova dicipiens	North Atlantic	cod
Gnathostoma sp.	Asia	freshwater fish, frogs
Capillaria sp.	Asia	freshwater fish
Angiostrongylus sp.	Asia, South America, Africa	freshwater prawns, snails, fish
Cestodes or tape worms		
Diphyllobothrium latum	Northern hemisphere	freshwater fish
D. pacificum	Peru, Chile, Japan	seawater fish
Trematodes or flukes		
Clonorchis sp.	Asia	freshwater fish, snails
Opisthorchis sp.	Asia	freshwater fish
Metagonimus yokagawai	Far East	
Heterophyes sp.	Middle East, Far East	snails, freshwater fish brackish water fish
Paragonimus sp.	Asia, America, Africa	snails, crustaceans, fishes
Echinostoma sp.	Asia	clams, freshwater fishes, snails

Another nematode of public health importance is *Capillaria* sp. (e.g. *Capillaria philippinensis*). The adult worms are gut parasites in piscivorous birds and intermediate hosts are small freshwater fish. Infection in human causes severe diarrhea and possible death attributed to fluid loss. A well known and common nematode in Asia is the *Angiostrongylus* sp. (e.g. *Angiostrongylus cantonensis*). The adult worm is found in the lungs of rats and the inter-mediate hosts are snails, freshwater prawns and land crabs. The parasite has been shown to cause meningitis in man (Figure 3.7).

Figure 3.7. Angiostronglid life cycle. The life of *Angiostrongylus* sp. is depicted. Sexually dioecious nematodes mate and produce eggs that pass with the faeces or hatch in the intestine. *A.cantonensis* reaches maturity in the lungs and *A.costaricensis* reaches maturity in the intestine. The larvae migrate in moist places and may invade invertebrates, such as gastropods. Mammals may encounter infective larvae through the consumption of undercooked infected invertebrates or vegetables. In mammals, the larvae penetrate the intestine and migrate in the viscera. *A.cantonensis* migrates through the subarachnoid space and develops before migrating to the lungs. In humans, larvae do not migrate beyond the brain. *A.cantonensis* migrates in the viscera, muscles and skin before returning to the intestine of rats. In humans it continues to migrate until it dies. The life cycle of the gnathostomatids is similar; they seem to infect and migrate in almost any intermediate host, but mature only in one that provides the proper physiological signal (after Brier 1992).

Cestodes

Only few *cestodes* or tapeworms in man are known to be transmitted by fish. However, the broad fish tapeworm *Diphyllobothrium latum* is a common human parasite reaching up to 10 m or more in length in the intestinal tract of man. This parasite has a microcrustacean as first intermediate host and freshwater fish are required as second intermediate host (Figure 3.8). The related species (*D. pacificum*) is transmitted by marine fish and commonly occurs in coastal regions of Peru, Chile and Japan where raw fish preparations (ceviche, sushi and others) are common.

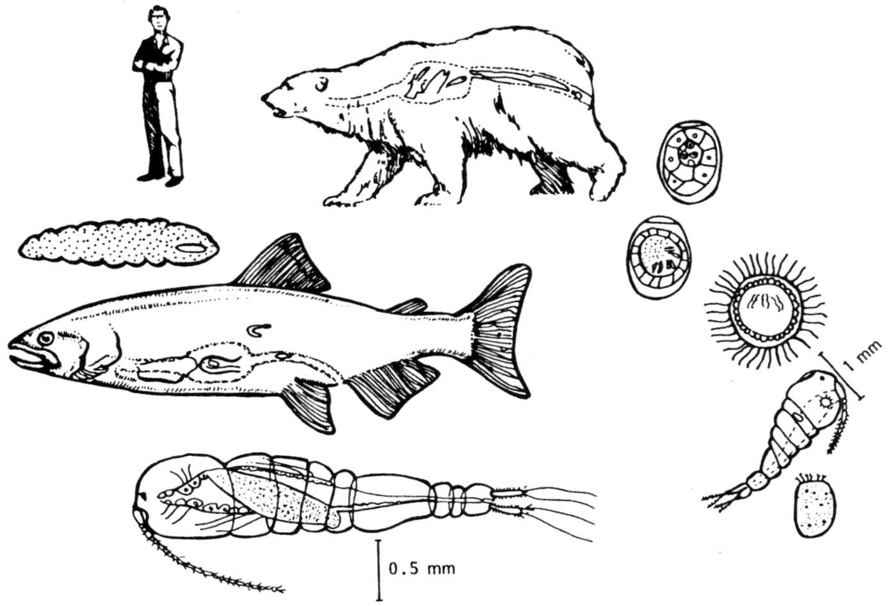

Figure 3.8. Broad fish tapeworm life cycle. The broad fish tapeworms, *Diphyllobothrium* sp., reach sexual maturity in the intestinal tract of mammals. Eggs may pass in the faeces and develop in water into larvae that hatch and swim freely. If consumed by a copepoda or other suitable crustacean host, the larvae may then become infective for fish that consume the infected crustacean. These larvae then develop into forms that may infect other fishes, where they do not develop further, or mammals, where they may reach sexual maturity (after Brier 1992).

Trematodes

Some of the *Trematodes* or flukes are extremely common, particularly in Asia. Thus it is estimated that the *Clonorchis sinensis* (the liver fluke) is infecting more than 20 million people in Asia. In Southern China human clonorchiasis rates can surpass 40% in some regions (Rim 1982). Intermediate hosts are snails and freshwater fish, while dogs, cats, wild animals and humans are final hosts where the fluke live and develop in the bile-ducts in the liver. The predominant problem in transmission is the contamination of snail-infested waters by egg-laden faeces from humans (e.g. use of "night soil" as fertilizers).

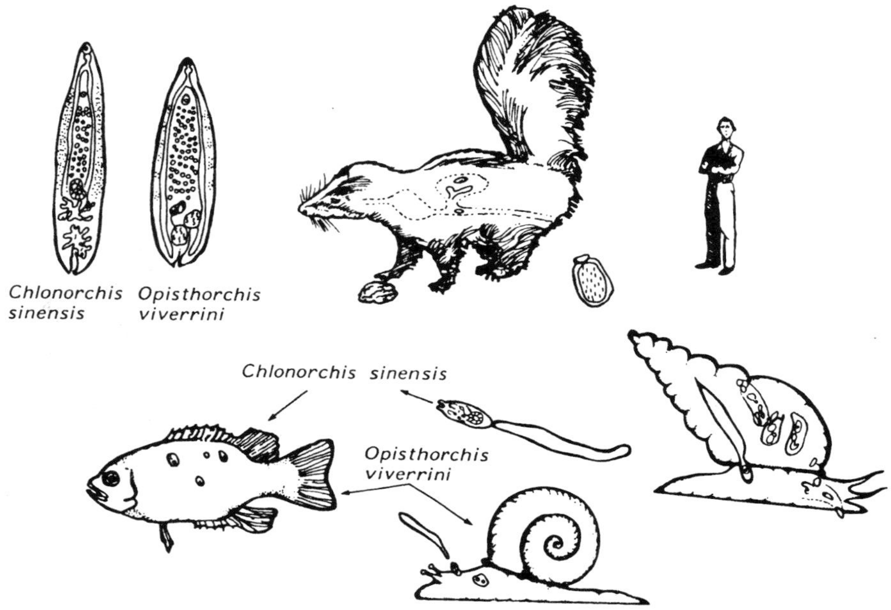

Figure 3.9. Liver fluke life cycle. These trematodes reach sexual maturity in the liver of humans and other mammals. Eggs enter the intestine in the bile, are incorporated into the faeces of the host and if ingested by a mollusc, may hatch. The larvae penetrate the tissues through morphologically distinct stages which by asexual reproduction produce free-swimming larvae. The larvae of *Clonorchis sinensis* can infect only certain fish species, whereas those of *Opisthorchis sinensis* can infect either fish or molluscan hosts. In these hosts the larvae become infective for mammals that consume raw or undercooked infected intermediate hosts (after Brier 1992).

Two very small flukes (1-2 mm) *Metagonimus yokagawai* and *Heterophyes heterophies* differ from *Clonorchis* by living in the intestines of the final host, causing inflammation, symptoms of diarrhea and abdominal pain. Intermediate hosts are snails and freshwater fish (Figure 3.10).

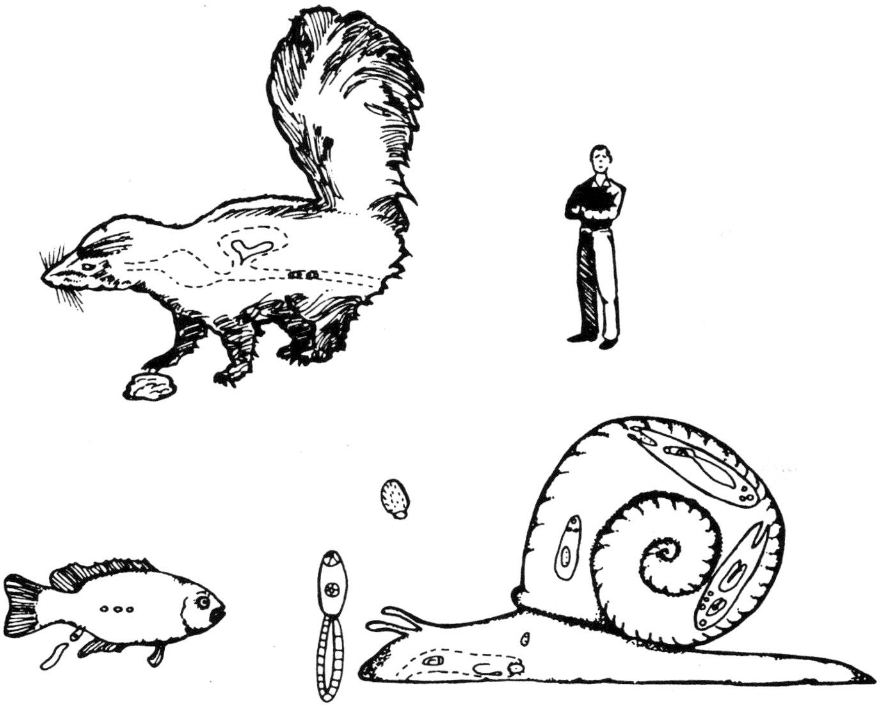

Figure 3.10. Heterophyid life cycle. Small intestinal flukes mature sexually in the small intestine of humans and other mammals. They mature deep in intestinal crypts, where some eggs may enter the circulatory system and cause cardiac damage. Eggs that exit in the faeces may develop into larvae, which, if consumed by a compatible gastropod host, hatch and penetrate into the snail's tissues where they develop through two morphologically distinct generations. Motile larvae that result leave the snail host and may penetrate into the tissues of a fish host to form the mammalian infective stage. The life cycle may be completed if humans or other mammals consume the infected fish hosts in a raw or undercooked state (after Brier 1992).

The adult oriental lung fluke *Paragonimus* sp. is 8-12 mm and encapsulated live in cysts in the lungs of man, cats, dogs and pigs and many wild carnivore animals. Snails and crustaceans (freshwater crab) are the intermediate hosts. (Figure 3.11).

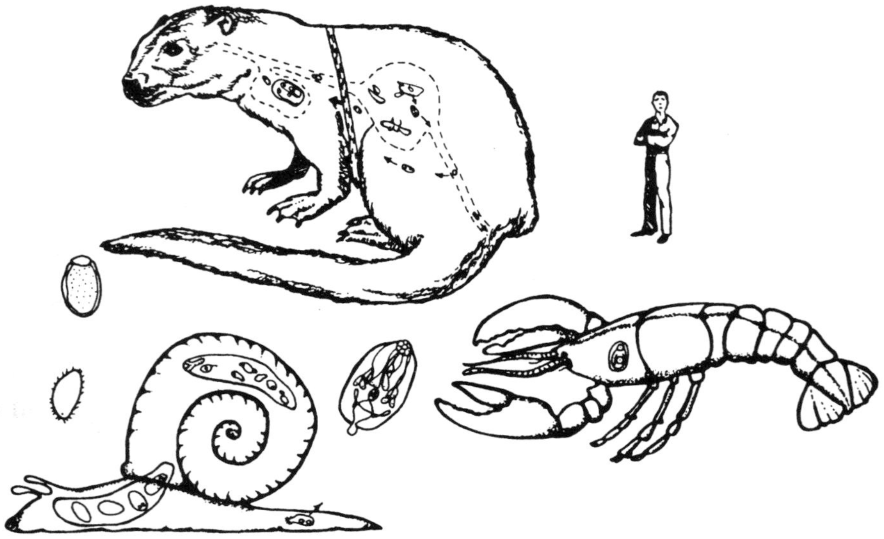

Figure 3.11. Lung fluke life cycle. *Paragonimus* sp. reach sexual maturity in the lungs of humans and other mammals and are usually found in pairs in the alveolar sacs. Eggs are coughed up and expelled by way of the sputum. They are also excreted in faeces. Free-living larvae hatch from the egg under suitably moist conditions. If these larvae encounter a gastropod host they may enter and develop asexually through two distinct morphological forms into free-living larvae, which then penetrate the soft tissues of a crab or crayfish and encapsulate as the mammalian infective stage. Larvae that are consumed by a mammal then penetrate the intestinal wall and migrate through the tissues. In some hosts, migration continues without further development; however, these larvae remain infective to mammals that consume the uncooked hosts. In hosts that provide the proper physiological signal, the larvae migrate to the lungs and mature (after Brier 1992).

Control of disease caused by parasites

All parasites of concern are transmitted to man by eating raw or uncooked fish products. Control measures to reduce the public health problem related to presence of parasites include legislation and surveillance. In principle the problem can be attacked at 3 levels as listed for nematodes by WHO (1989):

1. Avoidance of capture of nematode-infected fish by selecting specific fishing grounds, specific species or specific age groups.

2. Sorting and removal of nematode-infected fish or removal of nematodes from fish, e.g. by hand over a candling table.

3. Application of techniques to kill nematodes in the fish flesh.

Only 2) and 3) are applied in commercial fishery.

Control measures are particularly important for fish products, which are to be eaten raw or uncooked (matjes-herring, marinated fish, lightly salted fish and cold smoked fish, ceviche, sashimi, sushi etc.). Thus many national health regulations e.g. the German ordinance on health requirements for fish and shellfish (German Fish Ordinance 1988) contain specific rules for handling and processing this type of fish in order to make sure that all nematodes are killed (processing for safety). Based on coordinated research in Holland, Germany and Denmark (Huss et al. 1992), the following criteria for safe processing can be given:

Marinated fish:

Safe processing is primarily based on the level of NaCl in the tissue fluid. When the minimum amount of acetic acid (2.5-3.0% in the tissue fluid) is used, the following maximum, survival time of nematodes at various NaCl-levels was found:

% NaCl in tissue fluid	Max. survival time of nematodes
4-5	6>17 weeks
6-7	10-12 weeks
8-9	5-6 weeks

Maximum survival times of nematodes should therefore also be minimum holding time of the final product before sale.

Heat treated fish:

All nematodes were killed when heated to 55°C for 1 minute. This means that hot-smoked, pasteurized, sous-vide cooked and other lightly heat treated fish products are safe. However, some normal household cooking traditions may be on the borderline of safety.

Frozen fish:

Freezing to -20°C and maintaining of this temperature for at least 24 hours will kill all nematodes.

The results listed above show that a number of fish products are unsafe. This applies to lightly salted fish products (<5-6% NaCl in water phase) such as matjes-herring, gravad fish, cold smoked fish, lightly salted caviar, ceviche and several other local traditional products. A short period of freezing either of the raw material or the final product therefore must be included in the processing as a means of control of parasites.

3.6. CHEMICALS

Contamination with chemicals figure very low in official statistics as cause of seafood borne diseases (see Table 2.2).

The chemicals contaminants with some potential for toxicity appear to be (Ahmed 1991):

- Inorganic chemicals: antimony, arsenic, cadmium, lead, mercury, selenium, sulfites (used in shrimp processing).

- Organic compounds: polychlorinated biphenyls, dioxins, insecticides (chlorinated hydrocarbons).

- Processing related compounds: nitrosamines and contaminants related to aquaculture (antibiotics, hormones).

A modest concentration of contaminants are ubiquitous in the clean aquatic environment. A few metals such as copper, selenium, iron and zinc are essential nutrients for fish and shellfish. Contamination occurs when there is a statistically significant increase in the mean levels in comparable organisms.

Problems related to chemical contamination of the environment are nearly all man-made. The ocean dumping of hundreds of millions tons of waste material from industrial processing, sludge from sewage treatment plants, draining into the sea of chemicals used in agriculture and raw untreated sewage from large urban populations and industries all participate in contaminating the coastal marine environments or freshwater environments. From here the chemicals find their way into fish and other aquatic organisms. Increasing amounts of chemicals may be found in predatory species as a result of biomagnification, which is the concentration of the chemicals in the higher levels of the food chain. Or they may be there as a result of bioaccumulation, when increasing concentrations of chemicals in the body tissues accumulated over the life span of the individual. In this case, a large (i.e. an older) fish will have a higher content of the chemical concerned than a small (younger) fish of the same species. The presence of chemical contaminants in seafood is therefore highly dependent on geographic location, species and fish size, feeding patterns, solubility of chemicals and their persistence in the environment.

In a recent review on chemical residue concerns in seafood, Price (1992), concluded that risk from chemical contaminants in commercially harvested fish and shellfish is low and not a problem. Risk from chemical residues (mercury, selenium, dioxins, PCPs, kepone, chlordane, dieldrine and DDT) are primarily a concern with sport caught fish and shellfish, caught in coastal waters and (possibly) in highly polluted waters.

Nevertheless, a large section of a committee report concerned with Seafood Safety in U.S. (Ahmed 1991) has been devoted to occurrence of chemical contamination and related health risks. Some of the general conclusions and recommendations from this report are cited below:

- From both natural and human sources, a small proportion of seafood is contaminated with appreciable concentrations of potentially hazardous organic and inorganic chemicals. Some of the risks that may be significant include reproductive effects from PCBs and methylmercury, and carcinogenesis from selected PCB congeners, dioxins, and some chlorinated hydrocarbon pesticides.

- Consumption of some types of contaminated seafood poses enough risk that efforts toward evaluation, education and control of that risk must be improved.

- Present quantitative risk assessment procedures used by government agencies can and should be improved and extended to non-cancer effects.

- Current monitoring and surveillance programs provide an inadequate representation of the presence of contaminants in edible portions of domestic and imported seafood, resulting in serious difficulties in assessing both risks and specific opportunities for control.

- Because of the unevenness of contamination among species and geographic areas, it is feasible to narrowly target control efforts and still achieve meaningful reductions in exposures.

- The data base for evaluating the safety of certain chemicals that find their way into seafood via aquaculture and processing is too weak to support a conclusion that these products are being effectively controlled.

The principal recommendations of the committee are as follows:

- Existing regulations to minimize chemical and biological contamination of the aquatic environment should be strengthened and enforced.

- Existing FDA and state regulations should be strengthened and enforced to reduce the human consumption of aquatic organisms with relatively high contaminant levels (e.g. certain species from the Great Lakes with high levels of PCBs, swordfish and other species with high methylmercury levels).

- Federal agencies should actively support further research to determine the actual risks from the consumption of contaminants associated with seafood and to develop specific approaches for decreasing these risks.

- Increased environmental monitoring should be initiated at the state level, as part of an overall federal exposure management system.

- States should continue to be responsible for site closures, and for issuing health and contamination advisories tailored to the specific consumption habits, reproductive or other special risks, and information sources of specific groups of consumers.

- There should be an expanded program of public education on specific chemical contaminant hazards via governmental agencies and the health professions.

Some examples of maximum residual chemical contaminants in fish per human consumption are shown in Table 3.11.

Table 3.11. Examples of maximum residual chemical contaminants in fish for human consumption.

Chemical	Maximum residue limit (mg/kg)	Country
DDT + DDE + DDD	2	Denmark
Dieldrin	0.1	Sweden
PCB	2	Sweden
Lead	2	Denmark
Mercury	0.5	EEC

3.7. SPOILAGE

Fish tissue is characteristic in being rich in protein and non-protein - nitrogen (e.g. amino acids, trimethylamine-oxide (TMAO), creatinine), but low in carbohydrate resulting in a high post mortem pH (>6.0). Further, the pelagic, fatty fishes have a high content of lipids consisting mainly of triglycerides with long-chain fatty acids which are highly unsaturated. Also the phospholipids are highly unsaturated and these circumstances have important consequences for spoilage processes under aerobic storage conditions.

The condition named "spoilage" is not clearly defined in objective terms. Obvious signs of spoilage are:

- detection of off-odours and off-flavours
- slime formation
- gas production
- discolouration
- changes in texture

and the development of these spoilage conditions in fish and fish products is due to a combination of microbiological, chemical and autolytic phenomena.

Microbiological spoilage

Initial loss of quality of fresh (non-preserved) lean or non-fatty fish species, chilled or not chilled, is caused by autolytic changes, while spoilage is mainly due to the action of bacteria (see Figure 3.12).

The initial flora on fish is very diverse, although most often dominated by Gram-negative psychrotrophic bacteria. Fish caught in tropical areas, may carry a slightly higher load of Gram-positive organisms and enteric bacteria. During storage a characteristic flora develops, but only a part of this flora contribute to spoilage (see Table 3.12). The specific spoilage organisms (SSO) are producers of the metabolites responsible for the off odours and off flavours associated with spoilage.

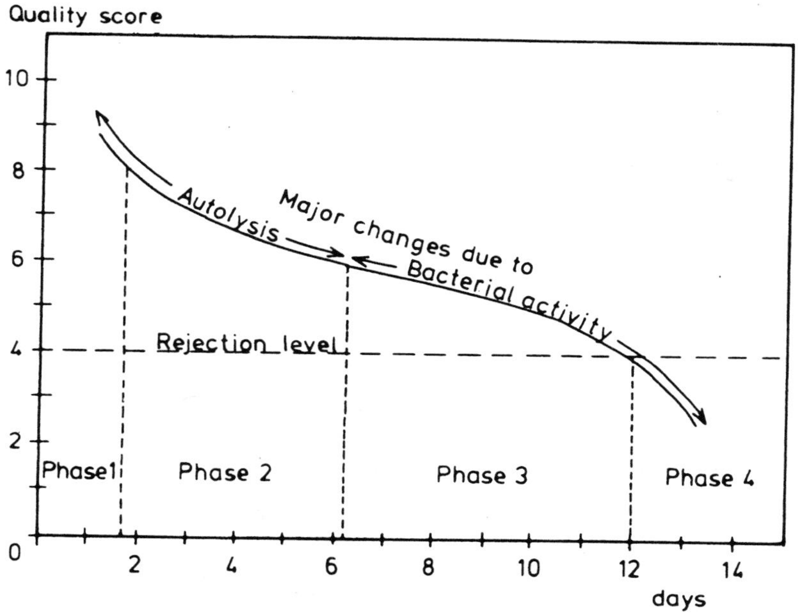

Figure 3.12. Changes in sensory quality of iced cod (0°C) (after Huss 1988).

Shewanella putrefaciens is typical for the aerobic chill spoilage of many fish from temperate waters and produces trimethylamine (TMA), hydrogen sulphide (H_2S) and other volatile sulphides which give rise to the fishy, sulphidy cabbage like off-odours and -flavours. Similar metabolites are formed by *Vibrionaceae* and *Enterobacteriaceae* during spoilage at higher temperatures. During storage in modified atmosphere (CO_2-containing), a psychrophilic *Photobacterium* producing large amounts of TMA is one of the major spoilage bacteria. Some fresh water fish and many fish from tropical waters are during iced, aerobic storage characterized by a *Pseudomonas* type of spoilage which is described as fruity, sulphydryl and sickening. Several volatile sulphides (e.g. methylmercaptan (CH_3SH) and dimethylsulphide ($(CH_3)_2S$), ketones, esters and aldehydes but not hydrogen sulphide are produced by *Pseudomonas* as are several ketones, esters and aldehydes. For fresh, non-preserved fish, the SSO which have been identified are shown in Table 3.12. Putrefaction or spoilage proceeds very rapidly once the load of SSO exceeds approximately 10^7 CFU/g.

Microbiological activity is also the cause of spoilage of many preserved fish products stored at temperatures $>0°C$. However, in most cases the specific spoilage bacteria are not known. The addition of small amounts of salt and acid, as in lightly preserved fish products, changes the dominating microflora to consist mainly of Gram positive bacterial species (Lactic acid bacteria, *Brochotrix*) and some of these may act as SSO under certain conditions as shown in Table 3.13. However, also some *Enterobacteriaceae* and *Vibrionaceae* may act as SSO for these products. In products with low levels of preservation, *Shewanella putrefaciens* may also play a role.

Also more strongly preserved fish products such as salt cured or fermented products spoil due to the action of certain microorganisms. The dominating flora on these products are Gram positive, halophilic or halotolerant micrococci, yeasts, spore formers, lactic acid bacteria and moulds. A number of SSO are known such as the extremely halophilic, anaerobic Gram negative rods and halophilic yeasts identified by Knøchel and Huss (1984) as specific spoilage organisms by causing off-odours and -flavours (sulphidy, fruity) in wet salted herring. An extreme halophile spoilage bacteria cause a condition known as "pink". These bacteria (*Halococcus* and *Halobacterium*) cause pink discolouration of salt, brines and salted fish as well as off odours and -flavours normally associated with spoilage (hydrogen sulphide and indole).

Some halophilic moulds (*Sporendonema*, *Oospora*) are also classified as spoilers. They do not produce off odours but their presence detracts from the value of the product because of their undesirable appearance.

Table 3.12. Dominating microflora and specific spoilage bacteria at spoilage of fresh, white fish (cod).

Storage temperature	Packaging atmosphere	Dominating microflora	Specific spoilage organisms (SSO)	References
0°C	Aerobic	Gram-negative psychrotrophic, non-fermentative rods (*Pseudomonas* sp., *S. putrefaciens*, *Moraxella*, *Acinetobacter*)	*S. putrefaciens* *Pseudomonas*[3]	2, 3, 4, 9
	Vacuum	Gram-negative rods; psychrotrophic or with psychrophilic character (*S. putrefaciens*, *Photobacterium*)	*S. putrefaciens* *P. phosphoreum*	1, 9
	MAP[1]	Gram-negative fermentative rods with psychrophilic character (*Photobacterium*) Gram-negative non-fermentative psychrotrophic rods (1-10% of flora; *Pseudomonas*, *S. putrefaciens*) Gram-positive rods (LAB[2])	*P. phosphoreum*	1, 7
5°C	Aerobic	Gram-negative psychrotrophic rods (*Vibrionaceae*, *S. putrefaciens*)	*Aeromonas* sp. *S. putrefaciens*	
	Vacuum	Gram-negative psychrotrophic rods (*Vibrionaceae*, *S. putrefaciens*)	*Aeromonas* sp. *S. putrefaciens*	
	MAP	Gram-negative psychrotrophic rods (*Vibrionaceae*)	*Aeromonas* sp.	6
20 - 30°C	Aerobic	Gram-negative mesophilic fermentative rods (*Vibrionaceae*, *Enterobacteriaceae*)	Motile *Aeromonas* sp. (*A. hydrophila*)	2, 4, 5, 8

1) Modified Atmosphere Packaging (CO_2 containing),
2) LAB: Lactic Acid Bacteria
3) Fish caught in tropical waters or fresh waters tend to have a spoilage dominated by *Pseudomonas* sp.

References: 1) Dalgaard et al. (1993), 2) Gram et al. (1987), 3) Lima dos Santos (1978), 4) Gram et al. (1990), 5) Gorczyca and Pek Poh Len (1985), 6) Donald and Gibson (1992), 7) van Spreekens (1977), 8) Barile et al. (1985), 9) Jørgensen and Huss (1989).

Table 3.13. Spoilage of lightly preserved fish products (salt content in water phase 3 - 6%, pH $\geq$ 5, temperature $\leq$ 5°C).

Product	Packaging atmosphere	Other preservatives than NaCl	Signs of spoilage	Dominating microflora	Specific spoilage organisms (SSO)[1]
Cold smoked fish	Vacuum	-	Off-odour / off-flavour (putrid, sickly, sulphurous)	Gram-negative rods (*Enterobacteriaceae*, *Vibrionaceae*) occasionally LAB[2]	???
			Off-flavour (sourness, acrid)	LAB	???
			Loss of aroma	LAB	-
Shrimps	In brine	Benzoic acid and/or sorbic acid; citric acid; pH 5.5 - 5.8	Slime	LAB	*Leuconostoc* sp.
			Gas production occasionally yeasty off-odour / off-flavour	LAB	Heterofermentative LAB, occasionally yeasts
			Diacetyl	LAB	LAB
		-	Off-odour / off-flavour	LAB, *Brochothrix*	???
Sugar-salted ('gravad') fish	Vacuum	-	Off-odour / off-flavour * Mackerel: rancid * Salmon: sour, acrid * Greenland Halibut: putrid	LAB, *Brochothrix*, occasionally Gram-negative bacteria (*Enterobacteriaceae*, *Vibrionaceae*, *S. putrefaciens*)	???
	MAP	-	Off-odour / off-flavour (sour)	Gram-positive bacteria (LAB)	???

1) i.e. specific spoilage organisms which have been related to the spoilage of the product
2) LAB = Lactic Acid Bacteria

Chemical spoilage (Oxidation)

The most important chemical spoilage processes are changes taking place in the lipid fraction of the fish. Oxidative processes, autoxidation, is a reaction involving only oxygen and unsaturated lipid. At first step leads to formation of hydroperoxides, which are tasteless but can cause brown and yellow discolouration of the fish tissue. The degradation of hydro-peroxides gives rise to formation of aldehydes and ketones as shown in Figure 3.13. These compounds have a strong rancid flavour. Oxidation may be initiated and accelerated by heat, light (especially UV-light) and several organic and inorganic substances (e.g. Cu and Fe). Also a number of antioxidants with the opposite effect are known (alpha-tocopherol, ascorbic acid, citric acid, carotenoids).

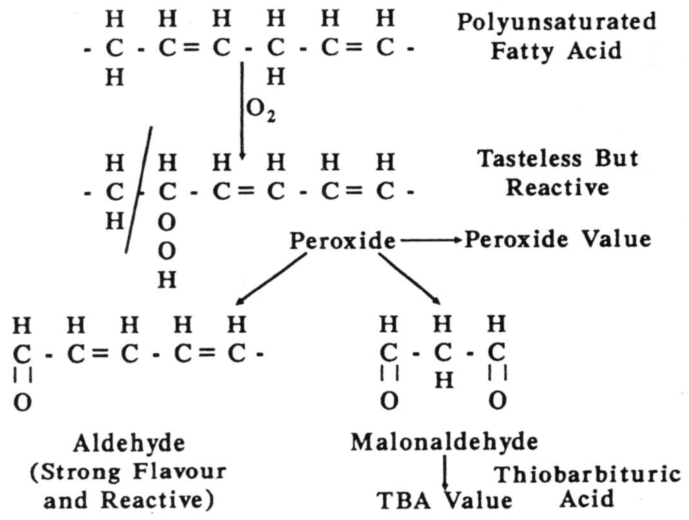

Figure 3.13. Basic processes for oxidation of polyunsaturated fatty acids found in fish tissue (after Ackman and Ratnayake 1992).

Autolytic spoilage

Autolytic spoilage or autolytic changes are responsible for early quality loss in fresh fish but contribute very little to spoilage of chilled fish and fish products. An exception from this statement is the rapid development of off-odours and discolourations due to action of gut enzymes in certain ungutted fish. However, in frozen fish the autolytic changes are of great importance. One example is the reduction of trimethylamine-oxide (TMAO), which in chilled fish is a bacterial process with formation of trimethylamine (TMA). In frozen fish, however, bacterial action is inhibited and TMAO is broken down by autolytic enzymes to dimethylamine (DMA) and formaldehyde (FA):

$$(CH_3)N{:}O \rightarrow (CH_3)_2\ NH + HCHO$$

The effect of the FA formed in frozen fish is increased denaturation of fish tissue, changes in texture and loss of water binding capacity. Other enzymatic reactions such as formation of free fatty acids are also believed to greatly influence the sensory quality of frozen fish. Autolytic enzymes are active even at -20°C and below, but are proceeding at a much faster rate at high, sub-zero temperatures.

The causes of the various types of spoilage are summarized in Table 3.14.

Table 3.14. Causes of fish spoilage

Signs of spoilage	Causes of fish spoilage			
	Microbiological	Chemical (oxidation)	Autolytic	Physical
Off odours/ off flavours	+	+	+	-
Slime formation	+	-	-	-
Gas formation	+	-	-	-
Discolouration	(+)	+	+	+
Change of texture	(+)	-	+	+

Control of spoilage

All proteinaceous foods spoil sooner or later, but a number of measures can be taken to reduce spoilage rate. Greatest effect can be obtained by control of storage temperature. As stated, the major cause of spoilage is bacterial, and in the chill temperature range the growth pattern of psychrotrophic spoilage organisms can be described accurately by the square root relation as reviewed by Bremner et al. (1987). Thus when 0°C is used as a reference temperature, the relationship comparing growth (r) at any particular temperature with that at 0°C becomes:

$$\sqrt{r} = 1 + 0.1 \times t \qquad \text{where t is temperature in °C.}$$

This means, if e.g. storage temperature is 10°C, the growth of spoilage bacteria is 4 times faster than at 0°C ($\sqrt{r} = 1 + 0.1 \times 10$, r=4) and shelflife is reduced correspondingly.

Chemical spoilage or development of rancidity can be prevented by rapid catch handling on board and storage of products under anoxic conditions (vacuum packed or modified atmosphere packed). Use of antioxidants may be considered.

The effect of storage temperature on quality of frozen fish is also pronounced and spoilage rate is considerably reduced at temperatures below -20°C.

The effect of hygiene in control of spoilage varies depending on the type of contamination which may take place. Great effort to reduce the general contamination during catch handling on board did not lead to any significant delay in spoilage (Huss et al. 1974) as only a very small part of this general contamination is made up of specific spoilage bacteria. In contrast, hygiene measures to control contamination of fish and fish products with specific spoilage bacteria greatly influences spoilage rate and shelflife (Jørgensen et al. 1988).

4. TRADITIONAL MICROBIOLOGICAL QUALITY CONTROL

Traditionally, three principal means have been used by governmental agencies and food processors to control microorganisms in food as listed by ICMSF (1988). These are (a) education and training, (b) inspection of facilities and operations and (c) microbiological testing. These programmes have been directed toward developing an understanding of the causes and consequences of microbial contamination and to evaluate facilities, operations and adherence to good handling practices. Although these are essential parts in any food control programme, they have certain limitations and shortcomings. The rapid turnover of personnel means that education and training must be a continuing exercise, which is rarely the case. As far as inspection of facilities and operations is concerned, this is often carried out with reference to various guidelines such as codes of practice, food control laws etc. These documents often fail to indicate the relative importance of the various requirements, and often these requirements are stated in very unprecise terms such as "satisfactory", "adequate", "acceptable", "suitable", "if necessary" etc. This lack of specificity leaves the interpretation to the inspector, who may place too much emphasis on relatively unimportant matters and thus increase costs without reducing hazards.

Microbiological testing also has some limitations as a control option. These are constraints of time, as results are not available until several days after testing as well as difficulties related to sampling, analytical methods and the use of indicator organisms. These problems will be discussed in more detail below followed by the description of a modified approach aiming at a preventive quality assurance programme.

Estimation of bacterial numbers in food is frequently used in the retrospective assessment of microbiological quality or to assess the presumptive "safety" of foods. This procedure requires that samples are taken of the food, microbiological tests or analyses are performed and the results evaluated - possibly by comparing with already established microbiological criteria. There are serious problems related to all steps in these procedures.

4.1. SAMPLING

The number, size and nature of the samples taken for analysis greatly influence the results. In some instances it is possible for the analytical sample to be truly representative of the "lot" sampled. This applies to liquids such as milk and water that can be sufficiently well mixed.

In cases of "lots" or "batches" of food this is not the case since a lot may easily consist of units with wide differences in the microbiological quality. A number of factors therefore must be considered before choosing a sampling plan (ICMSF 1986). These include:

- purpose of testing
- nature of product and lot to be sampled
- nature of analytical procedure.

A sampling plan (Attributes plan) can be based on positive or negative indications of a microorganism. Such a plan is described by the two figures "n" (number of sample units drawn) and "c" (maximum allowable number of positive results). In a 2-class attributes sampling plan, each sample unit is then classified into acceptable or non-acceptable. In some cases the presence of an organism (i.e. *Salmonella*) would be unacceptable. In other cases, a boundary is chosen, denoted by "m", which divides an acceptable count from an unacceptable. The 2-class sampling plan will reject a "lot" if more than "c" out of "n" samples tested are unacceptable.

In a 3-class sampling plan "m" separates acceptable counts from marginally acceptable counts and another figure "M" is indicating the boundary between marginally acceptable counts and unacceptable counts.

The safety which can be obtained with such sampling plans depends on the figures chosen for "c" and "n". This can be illustrated with the so-called operating characteristic curves which are demonstrating the statistical properties of such plans (Figure 4.1).

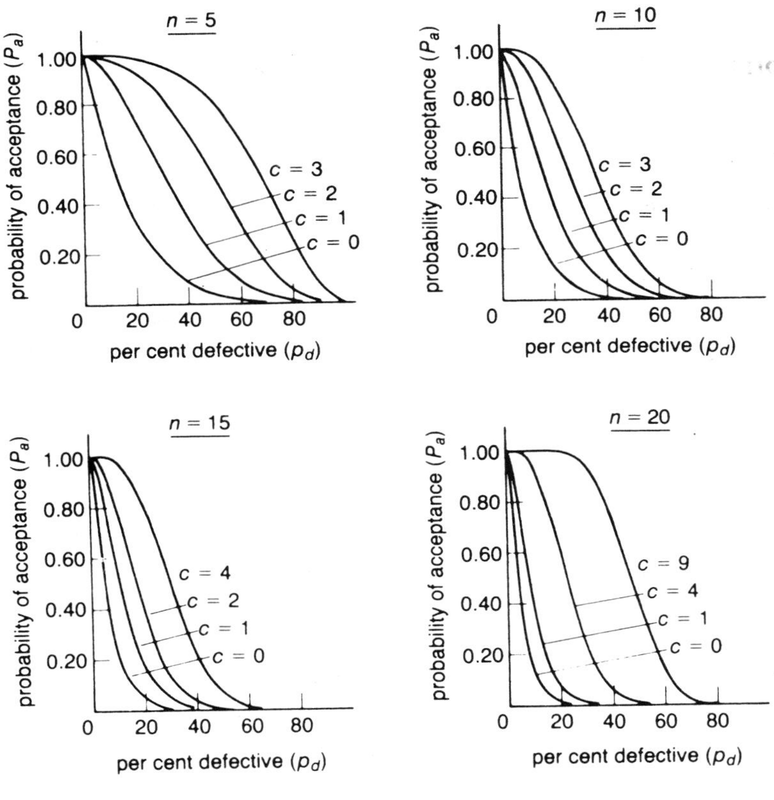

Figure 4.1. Operating characteristic curves for different sample sizes (n) and different criteria of acceptance (c) for 2-class attributes plan (ICMSF 1986).

Figure 4.1 shows that the greater the number of defective units (P_d), the lower is the probability of acceptance (P_a) of that lot. It is further demonstrated that high value of "n" and low value of "c" reduces the risk of accepting lots with same number of defective units. However, even the strictest sampling plans used are no great assurance of safety. Following those sampling plans recommended for infant formulas (n=60, c=o) involve testing 1.5 kg of food and even then there is a 30% risk of accepting product with 2% of sample units contaminated with *Salmonella*.

It is evident that even the most elaborate sampling of end-products cannot guarantee safety of the product.

It might be argued that although sampling and examination of samples may provide little assurance, it is still worthwhile in situations where there is no jurisdiction over handling and processing practices (such as for lots presented for acceptance at ports of entry). Even if only a fraction of the substandard consignments are found, the psychological effect on exporting companies is high.

In order to increase the relevance of sampling and testing, the International Commission on Microbiological Specifications for Foods (ICMSF) has introduced the concept of relating the stringency of the sampling plan to the degree of hazard of the food (ICMSF 1986). Thus the hazard may vary from a condition of no health hazard but only of utility (case 1-3) through low indirect health hazard (case 4-6) to moderate (case 7-12) and severe direct health hazards (case 13-15). In case of moderate or severe hazards, a 2-class attributes sampling plan is normally used. When the health hazard is low and in application of microbiological guidelines, a 3-class plan is suggested. For example, a typical 2-class plan with n = 5 and c = 0 requires that 5 sample units be tested and the lot would be rejected if one of the five sample units was defective. Table 4.1 shows the sampling plans and recommended microbiological limits suggested by ICMSF (1986) for seafood products.

The sampling plans applied by the Food and Drug Administration (FDA) for seafoods have been discussed and evaluated by a large Committee on Seafood Safety (Ahmed 1991). It is concluded that these sampling plans provide relatively little safety to the public and that increasing the sample size is not a reasonable solution. Even if testing methods for dangerous microorganisms, toxins and contaminant chemical were fully available and completely reliable, it is very clear that statistical uncertainties associated with lot sampling makes this an unreliable method for ensuring safety of food products. It is finally recommended by this Committee (Ahmed 1991) that suppliers of seafood to U.S. should be required to employ a Hazard Analysis Critical Control Point system, in order to obtain a high level of assurance and real-time control at the processing level.

Table 4.1. Sampling plan and recommended microbiological limits for seafood (ICMSF 1986).

Product	Test	Case	Plan Class	n	c	Limit per gram or per cm²	
						m	M
Fresh and frozen fish; cold smoked fish	APC[1]	1	3	5	3	$5x10^5$	10^7
	E.coli	4	3	5	3	11	500
Precooked breaded fish	APC	2	3	5	2	$5x10^5$	10^7
	E.coli	5	3	5	2	11	500
Frozen raw crustaceans	APC	1	3	5	3	10^6	10^7
	E.coli	4	3	5	3	11	500
Frozen cooked crustaceans	APC	2	3	5	2	$5x10^5$	10^7
	E.coli	5	3	5	2	11	500
	S.aureus	8	2	5	0	10^3	-
Cooked, chilled, and frozen crabmeat	APC	2	3	5	2	10^5	10^6
	E. coli	6	3	5	1	11	500
	S.aureus	9	2	5	0	10^3	-
Fresh and frozen bivalve molluscs	APC	3	2	5	0	$5x10^5$	-
	E. coli	6	2	5	0	16	-

1) APC = Aerobic Plate Count (preferably carried out at 21-25°C on a nutrient rich, non-selective agar.

4.2. MICROBIOLOGICAL TESTS

A number of microbiological tests of fish and fish products are used by industry for contractual and internal purposes and by authorities to check that the microbiological status is satisfactory. The purpose of these examinations is to detect for pathogenic bacteria (*Salmonella, V. parahaemolyticus, Staphylococcus aureus, Listeria monocytogenes, E. coli)* or for organisms which are possible indications of fecal contamination (*E. coli)* or other types of general contamination or poor manufacturing practices (coliform bacteria, fecal streptococci, aerobic plate count (APC).

Microbiological tests are generally costly, time-consuming and require a lot of manual labour but rapid automated tests are becoming available and being given accreditation. Consequently the number of samples which can be examined is limited. Furthermore, it should be emphasized again that a negative test for specific pathogens in a food sample is no guarantee that the whole lot is free of these pathogens. Thus only a very limited degree of safety can be obtained by microbiological testing. There are other limitations for some of these tests.

Total Viable Count (TVC) or Aerobic Plate Count (APC) is defined as the number of bacteria (cfu/g) in a food product obtained under optimal conditions of culturing. Thus the TVC is by no means a measure of the "total" bacterial population, but only a measure of the fraction of the microflora able to produce colonies in the medium used under the conditions of incubation. Thus it is well known that the temperature during incubation of plates influences greatly on the number of colonies developing from the same sample. As an example, the TVC may vary by a factor 10-100 when iced fish is sampled and plates are incubated at 20°C and 37°C respectively. Furthermore, the TVC does not differentiate between types of bacteria and similar levels of TVC may therefore be found although the biochemical activity of the bacteria may vary widely in the food. Also, high counts as a result of microbial growth are much more likely to cause defects in foods than are similar levels caused by recent gross contamination. TVC is therefore of no value in assessing the present state of sensory quality.

A TVC is meaningless as a quality index for products in group C and F (see Section 5.1.3) as a large population of non-spoilage lactic acid bacteria normally develop in these products. TVC is of very doubtful value in the examination of frozen fish products. An unknown and uncontrolled kill or damage of the bacteria may have taken place during freezing and cold storage. A very low "total" count may therefore lead to false conclusions about the hygienic quality of the product. Tests for TVC may be useful for measuring the conditions of the raw material, effectiveness of procedures (i.e. heat treatment) and hygiene conditions during processing, sanitary conditions of equipment and utensils and time x temperature profile during storage and distribution. However, to be useful and for correct interpretation of results a thorough knowledge of handling and processing conditions prior to sampling is essential.

E. coli: The natural habitat for this organism is the intestines of human and vertebrate animals. In temperate waters this organism is absent from fish and crustaceans at the time of capture (except in grossly polluted waters). Moreover, fish and shellfish should always be held at temperatures below those which support growth. This organism is therefore particularly useful as indicator of contamination (small numbers) or mishandling such as temperature abuse in product handling (large numbers). Contamination of food with *E. coli* implies a risk that one or more of enteric pathogens may have gained access to the food. However, failure to detect *E. coli* does not assure the absence of enteric pathogens (Mossel 1967, Silliker and Gabis 1976).

Recent investigations have shown that *E. coli* and fecal coliform bacteria can be found in unpolluted warm tropical waters and that *E. coli* can survive indefinitely in this environment (Hazen 1988, Fujioka et al. 1988, Toranzos et al. 1988). These studies also revealed that there was no correlation between presence or absence of fecal coliforms, total coliforms and virus. Thus, in the tropics *E. coli* or fecal coliforms are not reliable as indicators of recent biological contamination or sewage effluent discharge into the aquatic receptor. This point should be taken into consideration when microbiological standards are applied to fish products from tropical countries.

The resistance of *E. coli* to adverse physical and chemical conditions is low. This makes *E. coli* less useful as indicator organisms in examination of water and frozen or otherwise preserved fish products. Thus it is well established that enteric viruses survive much longer than *E. coli* in sea water (Melnick and Gerba 1980) and that *E. coli* is less resistant than *Salmonella* in frozen products (Mossel et al. 1980).

Fecal coliforms: This group of bacteria is often used in microbiological criteria instead of *E. coli* in order to avoid the lengthy and costly confirmation tests for *E. coli*. These organisms are selected by incubating an inoculum derived from a coliform enrichment broth at higher temperatures (44°C - 45.5°C). Thus, the group of fecal coliforms has a higher probability to contain organisms of fecal origin and hence indicating fecal contamination. Apart from being more rapid (- and less specific), a test for fecal coliform suffers the same limitations as described for *E. coli*. It should also be observed that the recently described pathogenic *E. coli* 0157:H7 does not grow at 44°C on all the selective media normally used for enumeration of *E. coli* (see Section 3.1.2).

Fecal streptococci or enterococci: It is now well established that fecal streptococci are not a reliable index of fecal contamination. Many foods and fish products contain these organisms as a normal part of their flora, and they are also able to establish themselves and persist in a food processing plant. Most are salt tolerant and may grow at 45°C as well as at chill temperatures (7-10°C). Unlike *E. coli*, they are relatively resistant to freezing, which makes them potentially useful as indicator organisms for evaluating plant hygiene during processing of frozen food.

Staphylococcus aureus: This organism is included in a number of microbiological criteria. Enumeration of this organism presents no problem. Spread plating on Baird-Parkers egg yolk medium and incubation for 30 hours at 37°C is a most reliable method. Positive cultures need to be confirmed by testing for coagulase activity.

The natural reservoir for *S. aureus* is human skin, hair and superficial mucous membranes (nose), while it is not a part of the normal flora on fish and fish product. Presence of large numbers indicate the possible presence of enterotoxin and/or faulty sanitary or production practice. Small numbers are to be expected in products handled by humans. It should be emphasized that *S. aureus* grows poorly in competition with large numbers of other microorganisms. For this reason, a test for *S. aureus* is only relevant for fish products which have received a bactericidal treatment, i.e. a heat treatment during processing. A test for toxin should be included if growth of *S. aureus* is suspected.

4.3. MICROBIOLOGICAL CRITERIA

A microbiological criterion is a standard, against which comparison and evaluation of own data can be made. A microbiological criterion may have either mandatory or advisory status. The various types of criteria have been defined by a subcommittee on microbiological criteria established by the U.S. National Research Council (FNB/NRC 1985):

- A microbiological *standard* is a microbiological criterion that is part of a law or ordinance and is a mandatory criterion.

- A microbiological *guideline* is a criterion used to assess microbiological conditions during processing, distribution and marketing of foods. Hence it is mostly an advisory criterion.

- A microbiological *specification* is used in purchase agreements between buyer and vendor.

Microbiological criteria may be helpful in assessing the safety and shelf-life of foods, the adherence to established Good Manufacturing Practice (GMP) and the suitability of food for a particular purpose. The various criteria will therefore often include both values for pathogenic bacteria or their toxins and indicator organisms.

It was further recommended by the subcommittee (FNB/NRC 1985) that a microbiological *criterion* should include the following components:

- A statement describing the identity of the food to which the criterion applies.

- A statement of the contaminant of concern, i.e. the microorganism or group of microorganisms and/or its toxin or other agent.

- The analytical method to be used for the detection, enumeration or quantification of the contaminant of concern,.

- The sampling plan.

- The microbiological limits considered appropriate to the food and commensurate with the sampling plan used.

Microbiological criteria should only be established when there is a need for it, and when it can be shown to be effective and practical. A number of factors should be considered as listed in FNB/NRC (1985). These include evidence of a hazard, nature of the product and the associate microflora, effect of processing, the state in which the food is distributed, the manner in which it is ultimately prepared for consumption, and whether reliable and practical methods of detection are available at a reasonable cost. A microbiological *standard* should be considered only when:

- There is clear evidence that a problem exists between a food and outbreaks of food-borne diseases and that the standard will alleviate the problem.

- Exceeding the limits is evidence that the food containing decomposed ingredients or it is processed or stored under grossly poor conditions.

- There is no jurisdiction over processing and distribution practices (i.e. imported foods), the standard will eliminate a health risk and/or reject products produced under questionable conditions.

Microbiological *guidelines* or reference values (Mossel 1982) are established as a result of surveys carried out during processing in a number (8-10) of factories where GMP is applied. Initially, all significant details of GMP are checked by visual inspection, instrumental methods or bacteriological tests. When everything is found to be in order, at least 10 samples from every checkpoint from every factory are drawn and examined. Distribution curves of the data obtained are prepared and used as a basis for establishment of reference values as suggested by Mossel (1982) (see Figure 4.2).

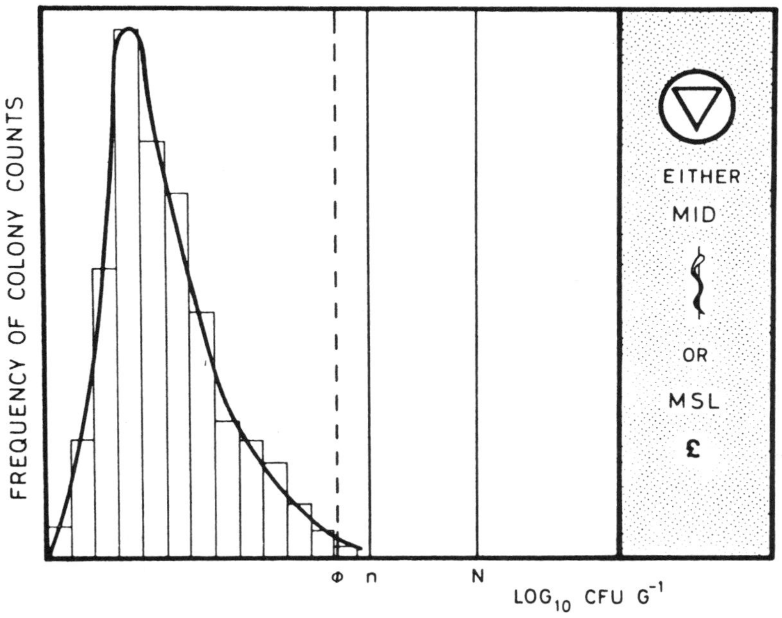

Figure 4.2. Distribution plot of the result of microbiological surveys on a given type of food (Mossel 1982).

Φ	- 95th percentile
n	- Reference Value proper
N	- maximal count to be expected under conditions of GMP
cfu	- colony-forming units
MID	- Minimal Infectious Dose
MSL	- Minimal Spoilage Level

The selection of values for n and N may vary according to the food involved and the local situation. As a general rule, N is one log cycle higher than n and one log cycle lower than MID or MSL. If Φ approaches MID or MSL too closely, improvement of the manufacturing technique is required. However, a certain tolerance has to be built into the reference values. The area between n and N is the "alert" area and the customary tolerance for *non-pathogenic* organisms is that no more than 2 out of 10 samples are expected in this range and none should show a cfu/g value over 10 times the reference value.

Microbiological guidelines are useful in determining the degree of control during processing and the conditions during distribution and storage. Thus the microbiological guideline can easily be incorporated in a HACCP-system (see Section 5.1) where they are useful as reference values in the monitoring work.

Also, microbiological *specifications* used in commercial transactions should be based on relevant background data and should fill a need. The currently used microbiological criteria applied to fish and fishery products by the members of the European Community together with Canada, Japan and USA (- these countries collectively import over 90% of the fish that is traded) are compiled by FAO (1989). The tests required are listed in Table 4.2.

It is apparent that the requirements of the microbiological criteria as listed above are not always considered in current practices for fish and fish products. Most of the standards listed in the FAO-circular (FAO 1989) are incomplete, unnecessary, unrealistic and should be reconsidered. In most cases only microbiological limits are specified and all the other components in a criterion are not considered. By careful evaluation of all aspects related to, for instance, fresh and frozen fish products which are intended to be heated before consumption, it is clear that these products do constitute neither a health risk nor a serious quality problem. The major problem related to these products is concerned with the possible presence of biotoxins. Thus, there is no need or justification for a microbiological criterion. Similarly, a large population of harmless lactic acid bacteria develop in lightly salted and cold smoked fish which makes a microbiological standard based on aerobic plate count (APC) meaningless. The inclusion of counts for *S. aureus* in microbiological standards for raw products with a large associate flora is also meaningless as already mentioned (Section 4.2).

A more realistic approach is taken by the ICMSF (1986) as shown in Table 4.1. Test for *S. aureus* is only recommended for cooked products, and *E. coli* is generally used as indicator for fecal contamination of all product types. However, the grouping of products in Table 4.1. is unscientific. Cold smoked fish are grouped with fresh and frozen fish although their microbiological ecology is vastly different, while frozen raw crustaceans form a group on its own although being microbiologically very similar to fresh and frozen fish. It is suggested that fish products should be grouped as shown in Section 5.1.3.

The microbiological limits recommended by ICMSF (1986) should be regarded as part of microbiological guidelines and useful mainly in the control of GMP. However, there is little or no evidence that these criteria have contributed significantly to the prevention of outbreaks of diseases attributed to these products. In view of the differences in the microbiological contamination of fish and crustaceans from various parts of the world, it is doubtful, whether such criteria are universally applicable.

Table 4.2. Microbiological tests included in the Microbiological Standards and Regulations of Some European Countries, Japan and USA. Belgium, Canada, Denmark, Germany, Greece and Portugal have no microbiological standards for fish and fish products. Data from FAO (1989).

	Italy	France	Luxembourg	Netherlands	United Kingdom	Spain	USA	Japan
Raw fish, fillets, fresh/frozen		1,2,7,10,11*)	1,3,7,10,11			1,2,5,6,7,10		1,2 (6)
Semi-preserves								
pasteurized		1,2,7,10,11						
non-pasteurized		1,2,7,10,11						
Smoked salmon		1,2,7,10,11						
Crustacean								
raw		1,3,7,11	1,3,7,11				1,6,10	
cooked		1,3,7,11	1,3,7,11		1,6,7,10			
cooked and peeled		1,3,7,10 11	1,3,7,10 11	7,10				
Molluscs								
live	6,7	3,4,7	3,4,7					
raw	6,7			6,7		1,6,7		1,6
pre-cooked	6,7	1,3,7,10,11	1,3,7,10,11			1,7,8,9,10		

*) The figures refer to tests for:

1. Aerobic plate count (TVC)
2. Coliforms
3. Fecal coliforms
4. Fecal streptococci
5. Enterococci
6. *E. coli*
7. *Salmonella*
8. *Shigella* sp.
9. Total enterobacteriaceae
10. *Staphylococcus aureus*
11. Anaerobic sulfite red.

In conclusion, it can be stated that there are no practical systems for providing safety or assurance of safety and normal shelf-life of fish products by microbiological end-product testing. Point-of-entry testing of fish products must generally be considered as an inefficient means of retrospective assessment of processing, transport and storage conditions. For this reason other methods should be used to assure a reasonable degree of protection of both consumer and producer against risks associated with microbiological activity. Apart from being useless as hygienic measures irrelevant criteria can still be of consequence by imposing unnecessary costs, introducing non-tariff barriers to trade and inducing a false sense of security.

However, microbiological criteria may be useful as a means for assessing the effectiveness of a quality assurance programme (HACCP) particularly as a part of a verification programme. This will be discussed in more detail in Section 5.1.3, but it cannot be overemphasized that microbiological criteria per se are totally inadequate.

By January 1st 1993 the single European market was established. The EEC Council Directive 91/493/EEC (EEC 1991b) is laying down the health conditions for the production and the placing on the market of fishery products. The directive gives provisions for laying down criteria for organoleptic quality, parasites, chemical checks (TVB-N, histamine and chemical contaminants) and microbiological analysis, including sampling plans and methods of analysis. So far, there are only criteria for histamine content of fish (9 samples must be taken from each batch. The mean value must not exceed 100 ppm, 2 samples may have a value of >100 ppm, but <200 ppm, no samples may have >200 ppm) and microbiological criteria for cooked, ready-to-eat shrimp and crabmeat, where the following *standards* apply:

1. *Salmonella* sp. - not to be detected in 25 g (n = 5, c = o)

2. *S. aureus* (cfu/g) m = 100, M = 1000 (n = 5, c = 2)

3. either thermotolerant coliforms (44°C) (cfu/g), m = 10, M = 100, (n=5, c=2) or *E. coli* (cfu/g), m = 10, M = 100, (n=5, c = 1).

See page 55 for explanation of n, c, m and M.

Furthermore the following microbiological *guidelines* apply for the same product.

Total viable count (aerobic, 30°C):

Whole product: m = 10,000, M = 100,000 (n=5, c=2)

Products without shell, not including crabmeat: m = 50,000 M = 500,000 (n=5, c=2)

Crabmeat: m = 100,000, M = 1,000,000 (n = 5, c=2).

For live, bivalve molluscs the requirements are listed in EEC-Council Directive 91/492/EEC of 15 July 1991 (EEC 1991a) as shown below:

Requirements concerning live bivalve molluscs

Live bivalve molluscs intended for immediate human consumption must comply with the following requirements:

1. The possession of visual characteristics associated with freshness and viability, including shells free of dirt, an adequate response to percussion, and normal amounts of intravalvular liquid.

2. They must contain less than 300 fecal coliforms or less than 230 *E. coli* per 100 g of flesh and intravalvular liquid based on a 5-tube 3-dilution MPN-test or any other bacteriological procedure of equivalent accuracy.

3. They must not contain *Salmonella* in 25 g of flesh.

4. They must not contain toxic or objectionable compounds occurring naturally or added to the environment.

5. The upper limit as regards the radionucleide contents must not exceed the limits for foodstuffs as laid down by the Community.

6. The total paralytic shellfish poison (PSP) content in the edible parts of molluscs must not exceed 80 μg per 100 g.

7. The customary biological testing methods must not give a positive result to the presence of diarrhetic shellfish poison (DSP) in the edible parts of the molluscs.

8. In the absence of routine virus testing procedures and the establishment of virological standards, health checks must be based on faecal bacteria counts.

Public health control

The public health control system must check, among other things, the microbiological quality of the live bivalve molluscs and the possible presence of toxin-producing plankton in the water and biotoxins in the molluscs. The sampling used for the control of toxins must be carried out in two steps:

1. Monitoring: Periodic sampling organized to detect changes in the composition of the plankton containing toxins and the geographical distribution thereof. Information leading to a suspicion of accumulation of toxins in mollusc flesh must be followed by:

2. Intensive sampling: The number of sampling points and the number of samples is increased and at the same time toxicity tests are introduced.

5. QUALITY ASSURANCE

While it is apparent that traditional quality control is unable to eliminate quality problems, a preventive strategy based on a thorough analysis of prevailing conditions is much more likely to provide assurance that objectives of the quality assurance programme are met. This point became very clear in the early days of food production and research for the U.S. space program (Bauman 1992). The amount of testing that had to be done to arrive at a reasonable decision point as to whether a food was acceptable in space travel was extremely high. Apart from the cost, a large part of any batch of food production had to be utilized for testing, leaving only a small portion available for the space flights. The result of these early considerations was the development of the Hazard Analysis Critical Control Point (HACCP) system, which was utilized by the Pillsbury Company's project on food production in the 60ies and exposed to the public during the 1971 National Conference on Food Protection (Anon. 1972).

The HACCP-system was and still is primarily aiming at guaranteeing food safety but can easily be extended to cover spoilage and economic fraud.

Further development and introduction of the HACCP-system into the general food production has been very slow (see Section 5.14). However, in recent years the system has been widely discussed and a number of new quality systems have been introduced, such as certification under an International Accepted Standard (ISO 9000 series) and Total Quality Management (TQM) in which everybody in an organisation is fully committed to achieving all aspects of quality.

One reason for this development is that a number of national food legislations today are placing full responsibility for food quality on the producer (EEC Council Directive 91/493/EEC (EEC 1991b)) and e.g. U.K.'s Food Safety Act (1990) is offering the possibility of a defense of due diligence being offered in prosecutions brought under the Act. This means that a truly documented system of quality assurance can support the plea that the manufacturer has been exercising due diligence. Harrigan (1993) stated that an organization may implement a quality system e.g. introduce HACCP, TQM or become certified under ISO 9001/2 for the following reasons:

- To improve the efficiency and profitability of their operations and the quality of their products.
- To satisfy a requirement from their customers/purchasers.
- To provide a due diligence defense in legal actions.
- To keep up with their competitors.

The advantage of having a documented, formal procedure for food quality assurance is now widely recognized. In the European Community (EC) a proposal for a Council Directive on the hygiene of foodstuff (EEC 1992) recognizes and requires the use of HACCP by food business operators while the application of standards of the EN 29000 series is recommended.

5.1. THE HAZARD ANALYSIS CRITICAL CONTROL POINT (HACCP)-SYSTEM

5.1.1. The HACCP-concept

After the basic principles of the system were published in 1971 (Anon. 1972) it has now been further elaborated by the ICMSF in publications for the World Health Organization (WHO) and finally in a book (ICMSF 1988). The HACCP-system has been widely discussed and unfortunately a number of new definitions and approaches have been published. This situation is likely to create some confusion and misunderstandings, unless some international agreements can be reached. A working group of the Codex Alimentarius Commission on Food Hygiene is presently (1992) drafting a report on HACCP, which hopefully will clarify these matters. However, the present publication will adhere very closely to the ICMSF definitions and strategy as outlined (ICSMF 1988). The system is based on the recognition that (microbiological) hazards exist at various points, but measures can be taken to control these hazards. The anticipation of hazards and the identification of control points are therefore key elements in HACCP. The system offers a rational and logical approach to control (microbiological) food hazards and avoid the many weaknesses inherent in the inspectional approach. Once established, the main effort of the quality assurance will be directed towards the Critical Control Points (CCPs) and away from endless final product testing. This will assure a much higher degree of safety - at less cost.

The main elements of the HACCP-system are:

A. Identify potential hazards.
Assess the risk (likelihood) of occurrence.

B. Determine the Critical Control Points (CCPs)
Determine steps that can be controlled to eliminate or minimize the hazards.

C. Establish the criteria (tolerances, target level) that must be met to ensure that CCP is under control.

D. Establish a monitoring system.

E. Establish the corrective action when CCP is not under control.

F. Establish procedures for verification.

G. Establish documentation and record keeping.

A. Identification of Potential Hazards

Hazards have been defined (ICMSF 1988) as the unacceptable contamination, growth or survival of bacteria in food that may affect food safety or quality (spoilage) or the unacceptable production or persistence in foods of substances such as toxins, enzymes or products of microbial metabolism. The U.S. National Advisory Committee on Microbiological Criteria for Foods (NACMCF 1992) have defined a hazard as: a biological, chemical or physical property that may cause a food to be unsafe for consumption (NACMCF 1992). For inclusion on the list, hazards must be of a nature such that their elimination or reduction to acceptable levels is **essential** to the production of safe food. (Some food companies also include regulatory compliance, nutritional value and other important aspects in the definition of hazards). Hazards which are of low risk and not likely to occur would not require further considerations (NACMCF 1992).

Thus, while the ICMSF includes both safety aspects and quality in the definition of hazards, the US-NACMCF is only including safety. In the present presentation, the HACCP-system will be used to control both safety and all aspects of spoilage of fish products.

Hazard analysis requires two essential ingredients. The first is an appreciation of the pathogenic organisms or any disease agent that could harm the consumer or cause spoilage of the product, and the second is a detailed understanding of how these hazards could arise. Thus the hazard analysis requires thorough microbiological knowledge in combination with epidemiological and technological information.

In order to be meaningful, hazard analysis must be quantitative. This requires an assessment of both **severity** and **risk**. Severity means the seriousness of the consequences when a hazard occurs, while risk is an estimate of the probability or likelihood of a hazard occurring. It is only the risk which can be controlled.

B. Determine the Critical Control Points (CCPs)

According to ICMSF, a CCP may be a location, procedure or processing step at which hazards can be controlled. Two types of CCPs may be identified: CCP-1 that will ensure full control of a hazard and CCP-2 that will minimize but not assure full control. Within the context of HACCP the meaning of "control" at a CCP means to minimize or prevent the risk of one or more hazards by taking specific preventative measures (PM).

According to the currently accepted definition by the US National Advisory Committee on Microbiological Criteria for Foods (NACMCF 1992) a CCP is: a point, step or procedure at which control can be applied and a food safety hazard can be prevented, eliminated or reduced to an acceptable level. (Note: No discrimination between CCP-1 and CCP-2). Thus for every step, location or procedure identified as a CCP, a detailed description of the preventative measures to be taken at that point must be provided. If there are no preventative measures to be taken at a certain point, it is not a CCP.

Thus CCPs should be carefully chosen on the basis of the risk and severity of the hazard to be controlled and the control points should be truly critical. In any operation many control points (CP) could be necessary but not critical due to low risk or low severity of the hazard involved. Some of these control points are there as a result of company rules for good manufacturing practice, product reputation, company policy or aesthetics. Such distinction between Control Points and Critical Control Points is one of the unique aspects of the HACCP-concept, which set priorities on risks and emphasizes operations that offer the greatest potential for control. Thus the HACCP points out what is **necessary** while further control may be **nice**.

It is not always easy to determine if a certain processing step is a CCP. A "decision tree" based on the ideas of Mayes (1992) and NACMCF (1992) as shown in Figure 5.1 may help to simplify this task. If an identified hazard has no preventive measure (PM) at a certain step then no CCP exists at that step and the question can be repeated at the next processing step. However, if PMs exist at the step it may or may not be a CCP depending on the step being specifically designed to eliminate the hazard under study.

Examples of CCPs are: A specified heat process, chilling, specific sanitation procedures, prevention of cross-contamination, adjustment of food to a given pH or NaCl content.

C. Establish criteria, target levels and tolerances for each CCP

To be effective, a detailed description of all CCPs is necessary. This includes determination of criteria and specified limits or characteristics of a physical (for example, time or temperature conditions), chemical (for example minimum NaCl concentration) or biological (sensory) nature which ensure that a product is safe and of acceptable quality. Establishment of safe criteria for utilizing a processing step (for example a heat treatment) as a CCP-1 for specified pathogens may necessitate extensive research work before implementation of the HACCP-system. Establishment of microbiological criteria (guidelines or reference values) at various processing steps or the final product also needs extensive investigation such as challenge studies or predictive modelling can be usefully applied when proper verified models are available (see also Section 4.3). Thus, a well equipped laboratory is needed for this purpose.

Other criteria such as moisture level, pH, a_w, chlorine level may be known from the technical literature. However, it must be emphasized that the HACCP-team must also define the processing conditions for obtaining safe food. Thus it is not sufficient to state that e.g. the internal temperature of a food must reach a certain temperature. The precise operation to obtain this target level using the available equipment must be determined, and the level of tolerance established. As examples: What is the maximum time at ambient temperature before icing, without significant quality loss? - or before any significant formation of histamine?

D. Establish a monitoring system for each CCP

The monitoring should measure accurately the chosen factors which control a CCP. It should be simple, give a quick result, be able to detect deviations from specifications or criteria (loss of control) and provide this information in time for corrective action to be taken. When it is not possible to monitor a critical limit on a continuous basis, it is necessary to establish that the monitoring interval will be reliable enough to indicate that the hazard is under control. Statistically designed data collection or sampling systems lend themselves to this purpose and frequency of measurements must be based upon the amount of risk that is acceptable to the management. The effectiveness of control should therefore preferably be monitored by visual observations or by chemical and physical testing. Microbiological methods have limitations in a HACCP-system, but they are very valuable as means of establishing and randomly verifying the effectiveness of control at CCPs (challenge tests, random testing, verification of hygiene and sanitation controls).

Apply HACCP Decision Tree to Each Step (Answer Questions in Sequence)

Q 1. Do Preventive Measures exist

YES NO Modify step, process or product.

Is control of this step necessary for safety? YES

NO NOT A CCP STOP

Q 2. Is the step specifically designed to eliminate or reduce the likely occurrence of a hazard to an acceptable level?

NO YES

Q 3. Does contamination with identified hazard(s) occur in excess of acceptable level(s) or could these increase to unacceptable level(s)?

YES NO NOT A CCP STOP

Q 4. Will a subsequent step eliminate identified hazard(s) or reduce likely occurrence to an acceptable level?

YES NO

NOT A CCP

STOP CRITICAL CONTROL POINT

Figure. 5.1. Decision tree to locate the Critical Control Points in a process flow (Mayes 1992, NACMCF 1992).

Record keeping and trend analysis constitute integral parts of monitoring as well as a reporting system. Monitoring records must be made available for review by regulatory authorities. All records must be signed by a designated person responsible for the quality aspects.

Since monitoring is a data collection activity, it is important to understand how to collect data. In general, there are ten steps to follow in designing a data collection (monitoring) activity (Hudak-Roos and Garrett 1992):

1. Ask the right questions. The questions must relate to the specific information needed. Otherwise, it is very easy to collect data that are incomplete or answer the wrong questions.

2. Conduct appropriate data analysis. What analysis must be done to get from raw data collection to a comparison with the critical limit?

3. Define "where" to collect.

4. Select an unbiased collector.

5. Understand the needs of the data collector, including special environment requirements, training, and experience.

6. Design simple but effective data collection forms. Remember "KISS" - keep it simple, stupid! Check to see that the forms are self-explanatory, record all appropriate data, and reduce opportunity for error.

7. Prepare instructions.

8. Test the forms and instructions and revise as necessary.

9. Train data collectors.

10. Audit the collection process and validate the results. Management should sign all data forms after review.

E. Corrective actions

The system must allow for corrective action to be taken immediately when the monitoring results indicate that a particular CCP is not under control. Action must be taken before deviation leads to a safety hazard. According to Tompkin (1992), corrective action involves four activities:

- Use the results of monitoring to adjust the process to maintain control.
- If control is lost, you must deal with non-compliance products.
- You must fix or correct the cause of non-compliance.
- Maintain records of the corrective actions.

It is important that **one** person is designated the responsibility to adjust the process and to inform others of what has happened. Tompkin (1992) also lists five options for dealing with non-compliance products:

- Release the product (not the wisest option if safety is involved).
- Test the product.
- Divert product to safe use.
- Reprocess the product.
- Destroy the product.

F. Verification

This is the use of supplementary information to check whether the HACCP system is working. Random sampling and analysis can be used. Other examples are the use of incubation tests for sterile or aseptically produced products, tests to see whether products can meet the expected and stated shelf-life and end product examinations. While frequent verifications using traditional microbiological methods may be applied, when the HACCP-system is first implemented, it can be reduced or even abolished when more experience is gained. Verifications could also be carried out by outside parties (government authorities, trade partners, consumer organizations, see also Section 5.1.4).

G. Establish record keeping and documentation

The approved HACCP-Plan and associated records must be on file. Documentation of HACCP procedures at all steps is essential. It should be clear at all times who is responsible for keeping the records. All documentation and records should be assembled in a manual and available for inspection by regulatory agencies.

5.1.2. Introduction and application of the HACCP-system

The principles of HACCP are all very logical, simple and straightforward. However, in the practical application a number of problems are likely to arise, particularly in large food factories. It is advisable therefore to adapt a logical and step-wise sequence for introduction of the HACCP-system as suggested by a HACCP-working group set up by the Codex Alimentarius, Food Hygiene Committee (Pierson and Corlett Jr. 1992), the International Association of Milk, Food and Environmental Sanitarians (IAMFES 1991), Mayes (1992) and Varnam and Evans (1991) as outlined below.

Step 1. Commitment

The first step is to ensure that top-management is firmly committed to introduce the system. Many departments and different personnel from chief to line operators will be involved and responsible for part of the system and their full support and cooperation will be essential. However, it is essential that one person (chief of quality assurance) is taking responsibility for the general and overall operation of the system.

Also sufficient resources (personnel, equipment) must be made available for implementation of HACCP.

Step 2. Assemble the HACCP-team and materials

Introduction of a HACCP-system in large food factories is a complex process and requires a multidisciplinary approach by a team of specialists. The microbiologist is of paramount importance, and must advise the team on all matters related to microbiology, safety and risks. He must have an updated knowledge on these matters and also access to technical literature on the most recent developments in his field. In many cases, he will also need access to the use of a well-equipped laboratory if specific questions and problems cannot be solved by studying the technical literature. Examples are investigations of the microbial ecology of specific products, challenge tests and inoculation studies for evaluation of safety aspects.

Another important member of the HACCP-team is the processing specialist. He must advise on production procedures and constraints, prepare the initial process-flow diagram, advise on technological objectives at various points in the process and on technical limitations of equipment.

Other technical specialists such as a chemist, quality assurance manager, the engineer as well as packaging technologists, sales staff, training and personnel managers can provide valuable information to the HACCP-team and they should attend some of the meetings.

Key-members of the HACCP-team (including the chairman) must have an intimate knowledge of the HACCP-system. Small and medium size industries are not likely to have qualified personnel on the payroll and must therefore buy assistance from outside consultants in order to implement the system.

Step 3. Initiation of program

When the HACCP-team is assembled, their terms of reference must be clearly defined and agreed to by the group. The work may be subdivided into a series of studies, each study dealing with a specific hazard (e.g. *C. botulinum* as a potential hazard in cold smoked salmon) or production of a specific product including all hazards related to this particular product. Whatever is decided, it should always be clear that the HACCP-system is unique and specific to each processing unit. The HACCP-concept is general but the application is specific to each and every local situation.

A detailed description and specification of the product must be provided to the team at this stage. The specification must include all technological aspects including preservative parameters (NaCl, pH, use of organic acids, other preservatives), intended storage temperature, packaging technology and - most important - intended use of the product. The intended processing technology ingredients list, a precise process flow diagram, description of cleaning and sanitation procedures must also be provided.

A visit to the processing site for verification and to fully understand the process-flow diagram may be made at this stage. Also the facility and equipment designs must be inspected to obtain information on the possibility of additional hazards related to these aspects (i.e. lay-out, traffic patterns for people, equipment properly sized for the volume of food to be processed etc.).

Step 4. Process analysis

When all the information regarding product and process has been collected, the data must be analyzed, all hazards identified and Critical Control Points (CCPs) established (HACCP elements A and B). To help in this process, a decision tree as shown in Figure 5.1 can be very helpful. Each step in the process must be analyzed separately and thoroughly, and the main questions must be asked and answered. This involves a discussion not only on the processing steps, but also on the intermediate stages between processing operations. As an example the time-temperature conditions during necessary holding periods can be mentioned.

The level of concern must be assessed at each processing step. This is to ensure that most efforts are directed towards the most critical areas. This level of concern can be assessed in various ways, but in most cases an expert judgement of risk based empirically on available data is sufficient. If this is not possible, tests or investigations may have to be carried out.

All true CCPs must be identified on the flow diagram. If other control points, which are not critical also are marked on the flow diagram, a clear distinction must be made.

Step 5. Control procedures

Each CCP must have a clear and specific control procedure which specifies how the CCP will be controlled. The preventive measure must be described in detail and target values and acceptable degree of latitude (if any) must be specified as well as when and how control measurements must be carried out (HACCP element C). In Table 5.1 some examples of control procedures are demonstrated.

Equipment and instruments used in control functions must also be kept under strict control, and their performance must be validated regularly.

Table 5.1. Examples of control procedures

Example of Hazard	Critical Control Point	Control procedures
Growth of *C. botulinum*	Salting of salmon to be smoked	Required salt level: 3-3,5% NaCl in water phase of fish Samples for testing from every batch
Contamination	Chlorination of cooling water	Continuous chlorine monitor, daily sampling of water for testing. Limit: 5 ppm, tolerance 3-5 ppm.
Contamination	Factory hygiene	Specification of cleaning and sanitation procedures. Visual control before work start. Twice weekly microbiological controls of cleaned surfaces in contact with food. Limit <100 cfu cm^{-2} Tolerance: Mean <100 cfu cm^{-2}. Max. 10^3 cfu.
Survival of pathogens	Cooking	Define and ensure time/temperature requirements. Continuous and automatic recording of water temperature

Step 6. Monitoring procedures

Monitoring and data recording are essential elements of the system. All actions, observations and measurements must be recorded for possible later use. These records are the tools by which management and outside inspectors are able to ensure that all operations are within specifications and that all CCPs are under full control. A high level of documentation - preferably confirmed by the signature of the controller - is also a sign of a high level of control.

Data not directly related to process control should also be recorded. Thus a detailed record of the initial HACCP-study including possible challenge tests or shelf life experiments should also be kept. In addition all changes to product-formulas or processing lines introduced as a result of the HACCP-study must be on record, as well as corrective actions taken when something was out of control.

Step 7. Training of staff

When the HACCP-study is completed and the programme is ready for implementation, training of staff must take place. All persons involved in the programme from line operators to managers must understand the principles and have a very clear idea of their own role in the system. Training and refresher courses must take place regularly and new staff should not be allowed to begin work before they have gone through training in HACCP principles and procedures.

On-going programme

The initial HACCP-study requires specific expertise in various fields, as well as access to a well equipped laboratory as already stated. In contrast, the daily routines in monitoring the system are quite simple and do not require e.g. expert microbiological skill and little or no laboratory expertise is needed. For small and medium size food companies it is therefore an economical advantage to buy outside expertise to introduce the system - and probably also to carry out the occasional verifications. The expense of installing costly laboratories, employment of highly paid skilled microbiologists - and the running cost of carrying out large numbers of unnecessary end-product analysis can be avoided.

Periodic reviews should be conducted to ensure that the HACCP-system is working correctly and that new developments are taken into account. Keypersonnel as well as line operators should be interviewed to assure that they understand their role in the programme. All changes in product or processing procedures should always be critically assessed before being instituted.

The principles of HACCP-system is equally applicable to large companies with complicated and large varieties of products and processing lines and to small enterprises, with a small production of one or a few simple product types. Naturally, the latter type of processing does not need a large HACCP-team and in-depth studies to introduce the system, as most of the answers are known beforehand. However, the advantage by the system of providing maximum quality assurance in the most cost effective way will also apply equally to both types of processing plants.

5.1.3. Use of the HACCP-concept in seafood processing

The final application of the HACCP-concept in any food processing is unique for every process and for every factory. In each case a detailed study of process flow is necessary in order to identify the hazards and the CCPs. However, some general principles can be outlined. For this purpose, seafood having similar microbiological ecology, handling and processing practice and/or similar culinary preparations before consumption can conveniently be grouped and categorized as shown below.

Seafood hazard categories:

A. Molluscs, including fresh and frozen mussels, clams, oysters in shell or shucked. Often eaten with no additional cooking.

B. Fish raw materials, fresh and frozen fish and crustaceans. Usually eaten after cooking.

C. Lightly preserved fish products (i.e. NaCl $<6\%$ (w/w) in water phase, pH >5.0). This group includes salted, marinated, cold smoked and gravad fish. Eaten without cooking.

D. Heat-processed (pasteurized, cooked, hot smoked) fish products and crustaceans (including pre-cooked, breaded fillets). Some products eaten with no additional cooking.

E. Heat-processed (sterilized, packed in sealed containers). Often eaten with no additional cooking.

F. Semi-preserved fish (i.e. NaCl $>6\%$ (w/w) in water phase, or pH <5.0, preservatives (sorbate, benzoate, NO_2 may be added). This group includes salted and/or marinated fish and caviar. Eaten without cooking.

G. Dried, dry-salted and smoke-dried fish. Usually eaten after cooking.

In ranking seafood in risk categories the method of NACMCF (1992) with some modifications has been applied. Thus a number of hazard characteristics are listed as shown below:

I There is epidemiological evidence that this type of product has (often) been associated with foodborne disease.

II The production process does not include a CCP-1 (i.e. full control) for an identified hazard.

III The product is subject to potentially harmful recontamination after processing and before packaging.

IV There is substantial potential for abusive handling in distribution or in consumer handling that could render the product harmful when consumed.

V There is no terminal heat process after packing or when cooked in the home.

The various seafoods can then be assigned to a risk category in terms of health hazards by using a + (plus) to indicate a potential risk related to the hazard characteristics. The number of pluses will then determine the risk category of the seafood concerned as shown in Table 5.2.

Table 5.2. Assignment of risk categories[1] for seafood

Seafood products	Hazard characteristics					Risk category
	I Bad safety record	II No CCP-1 for identified hazard in process	III Recontamination between processing and packaging.	IV Abusive handling during distribution and consumption	V No terminal heat process by consumer	
Molluscs (to be eaten raw)	+	+	+	+	+	High[1]
Fish raw materials Fresh and frozen fish and crustaceans	(+)[2]	+	-	-	-	Low
Lightly preserved	+	-	+	+	+	High
Heat-processed (pasteurized)	+	-	+	+	+	High
Heat-processed (sterilized)	(+)	-	(+)	-	+	Low
Semi-preserved	(+)[3]	-	-	(+)	+	Low
Dried, dry salted and smoke dried		-	-	-	(+)	No risk if cooked

1) High risk products have 3 or more pluses; Low risk products have less than 3 pluses.
2) Bad safety record for fish/frozen fish is mainly related to areas with possible presence of biotoxins.
3) Reported outbreak of botulism mostly due to toxin formation in raw material.

A. Molluscs

The molluscan shellfish are harvested by being raked or trawled from the bottom (oysters, mussels) or dug from the sand at low tide (clams and cookels). After harvesting, the shellfish are sorted (size), washed and packed in bags or crates or just left in a pile on deck. The shellfish may be transported and sold live to the consumer or they may be processed (shucked) raw and by use of heat. The heat applied in processing is only enough to facilitate shucking by causing the animal to relax the adductor muscle, and has no effect on the microbial contamination of the animals. The shucked meat is washed, packed and sold fresh, frozen or further processed and canned.

Most molluscs (oysters, mussels, clams, cockles) grow and are harvested in shallow, near-shore estuarine waters. Thus there is a strong possibility that the live animals may be contaminated with sewage-derived pathogens as well as those from the general environment. Due to the filter feeding of molluscs, a high concentration of disease agents may be present in the animals and therefore constitutes a serious hazard.

Most molluscs are traditionally eaten raw or very lightly cooked. They are therefore obviously a very high risk food, as also confirmed by the epidemiological evidence presented by Garret and Hudak-Roos (1991) who reported that 7% of all outbreaks of seafood-borne diseases (20% of all cases) in the U.S. in the period 1982-87 were caused by molluscan shellfish.

Table 5.3. Analysis of safety hazards in molluscs-processing.

Organism/component of concern	Hazard			
	Contamination	Growth	Severity	Risk
Pathogenic bacteria				
indigenous	+	+[1)]	high/low[2)]	high
non-indigenous	+	+[1)]	high	high
Virus	+	-	high/low[2)]	high
Biotoxins	+	-	high/low[2)]	high
Biogenic amines	-	-	-	-
Parasites	+	-	low	high
Chemicals	+	-	high/low[2)]	low
Spoilage bacteria	+	+	+	high

1) Growth of bacteria in molluscs after harvesting is only related to dead animals.
2) Severity of illness depends on organism or toxin involved.

Spoilage of dead molluscs is rapid, regardless of being shucked or not. However, the risk of contamination of the meat with specific spoilage bacteria is great during processing and packaging. A summary of hazards in molluscs-processing is shown in Table 5.3.

Unfortunately, it is not possible to control the many high risk hazards associated with consumption of raw molluscs. No CCP-1 can be identified for serious hazards such as contamination of live and dead animals with disease agents. These hazards can be reduced but not eliminated by means of :

- Control with the environment of live molluscs.
- Factory hygiene, including control of water quality.

This means, that these measures are only of CCP-2 nature.

The hazards related to growth of bacteria in the dead shellfish can be completely controlled by low temperature. Thus time and temperature conditions are CCP-1 for this particular hazard as shown in Table 5.4.

Controlling the environment of live molluscs

Growing and harvesting of molluscs should be restricted to areas free from direct sources of sewage pollution. This requires knowledge of local geography, prevailing water currents and the local treatment and discharge of sewage. Also monitoring of the microbiological quality of water is required. Thus the current standard for water quality in growing areas in U.S. is 14 MPN fecal coliforms/100 ml water, with no more than 10% of samples exceeding 43 MPN fecal coliforms/100 ml (FDA 1989). However, the value of fecal coliforms as indicator of contamination and possible presence of disease agents have serious limitations as already discussed (Section 4.2).

Also the correlation between presence of indicator bacteria and various disease agents in water and in shellfish has been questioned. The concentration of microorganisms in filter feeding shellfish varies enormously from animal to animal and also depends on weather conditions, temperatures and general activity of the shellfish. For these reasons, there are no microbiological standards for water quality in growing areas specified by the European Economic Community (EEC). Instead, the EEC has placed a microbiological standard (EEC Directive 91/492/EEC) on shellfish for direct consumption (EEC 1991a):

<300 fecal coliforms/100 g meat or
<230 *E.coli*/100 g meat (based on MPN-test)

absence of *Salmonella* in 25 g meat
PSP < 80 μg/100 g edible meat
DSP not to be detected by customary biological testing method

FDA (1989) also specifies a microbiological standard for shellfish as <230 MPN fecal coliform pr. 100 g meat and aerobic plate count (APC) of not more than 500,000/g meat.

Table 5.4. Safety hazards and preventive measures during processing and distribution of chilled molluscs.

Product flow	Hazard	Preventive measure	Degree of control
Live molluscs	Contaminated[1)]	Monitoring of environment	CCP-2
Catching			
Chilling	Growth of bacteria	(Txt) control[2)]	CCP-1
Transport	Growth of bacteria	(Txt) control	CCP-1
Reception at factory			
Shucking			
Packaging			
All processing steps	Growth of bacteria Contamination	(Txt) control Factory hygiene Water quality Sanitation	CCP-1 CCP-2 CCP-1 CCP-2
Chilling	Growth of bacteria	(Txt) control	CCP-1
Distribution	Growth of bacteria	(Txt) control	CCP-1

1) Hazards are contamination with pathogenic bacteria, virus, biotoxins, parasites and chemicals. All hazards except spoilage bacteria from Table 5.3.
2) (Txt) control = Time x temperature control

It should be noted here that the regulatory agencies in US and Europe have returned to traditional control options (sampling, testing and comparing results with microbiological standards) in an attempt to provide safety in consumption of raw molluscs. As already discussed in Section 4, these methods do not provide any guarantees for safety, and by the application of HACCP-concept it is also clear that no guarantee of safety can be provided. This point should be made clear to consumers, who insist on consumption of raw molluscs. Such warnings are actually used in seafood restaurants in Florida, USA.

Also control of the environment for the presence of toxic dinoflagellates is difficult and facing some of the same types of problems as discussed for bacteria and viruses. The EEC (EEC 1991a) requires periodical (weekly) sampling of water and shellfish from growing and harvesting areas, and in case of elevated presence of toxic algae, the fishing area will be closed. However, analytical techniques is one of the great problems.

An alternate means of securing safety of shellfish is by relaying or depuration and in a number of countries this is required by regulation. Depuration involves placing the shellfish in tanks with clean circulating seawater. Various methods can be used to disinfect the water such as ultra-violet light, chlorine, iodophors, ozone and activated oxygen (Richards 1991) and the shellfish is simply removed from suspect areas to waters which are known to be unpolluted. Both methods are of limited efficiency in the removal of viruses and vibrios from the shellfish (Richards 1991). Efficiency has normally been verified by testing the animals for presence of *E. coli*. This organism is not suitable as indicator, however, and alternative method is required. Non microbial contamination (toxins/biotoxins, heavy metals, petroleum, hydrocarbons, radionucleides, pesticides) depurate so slowly that commercial depuration is uneconomical (Richards 1991).

In most countries control and monitoring of the environment is the responsibility of the governments and these should be consulted for detailed information. It is expected that governments will ban fishing or harvesting of molluscs, if criteria are not met.

Temperature control

Time-temperature (Txt) conditions at all times from catching to distribution is a CCP-1 in preventing growth of pathogens and spoilage bacteria. Thus the time-lapse between each step in the flow diagram (Table 5.4) must be monitored and similarly the temperature of the environment, chill rooms, factory etc. as well as the temperature of the product must be recorded.

Factory hygiene and sanitation

Factory hygiene as well as personal hygiene and sanitation are CCPs in the prevention of contamination of products with microorganisms, filth and any other foreign material during processing. The seriousness (risk) of this hazard varies depending on local conditions (factory lay-out and design, facilities) and intended use of product (cooking or no cooking before being eaten). For this reason, a detailed description of the requirements must be produced in each individual case. These instructions must specify precisely when to clean and sanitize, how to do it, who is responsible, equipment and chemical agents to be used etc. (see also Section 6).

This CCP can then be controlled and monitored by visual inspection of procedures and recording of data in check-lists as shown in the example in Tables 5.7 and 5.8.

Occasionally a microbiological check of cleanliness of surfaces coming into direct contact with exposed meat can be made. Bacteriological control must be regarded more as a verification procedure than monitoring of a CCP. The frequency to carry out this type of check also depends on circumstances. In cases where changes in procedures or personnel has taken place. This control procedure must be carried out on a weekly or maybe on a daily basis. In other cases where routines are well established, microbiological control of cleanliness can be carried out monthly or maybe abolished altogether.

Water quality is a CCP-1 in preventing contamination from this source. Monitoring of this CCP-1 can be carried out by microbiological testing. Where in-plant chlorination is used, chlorine levels must be measured and recorded. Chlorine levels should be measured daily with recommended levels of 2-5 ppm.

B. Fish raw materials for further processing. Fresh and frozen raw fish and crustaceans

The hazard analysis of these products is fairly straightforward and uncomplicated. The live animals are caught in the sea or freshwater, handled and - in most cases - processed without any use of additives or chemical preservatives and finally distributed with chilling or freezing as the only means of preservation. Most fish and crustaceans are cooked before eating although a few countries such as Japan have a tradition for eating raw fish. The epidemiological records show that these products have caused a number of food poisoning outbreaks, but nearly all have been related to the presence of heat stable toxins (biotoxins, histamine).

Live fish and crustaceans and raw products may be contaminated with a number of pathogenic bacteria normally found in the aquatic environment such as *C. botulinum*, *V. parahaemolyticus,* various *Vibrio* sp., *L. monocytogenes, Aeromonas* sp. However, only the growth of these organisms can be regarded as a hazard, as pathogenicity is related to preformed toxin in the food (*C. botulinum*) or the minimal infection dose is known to be high (*Vibrio*). The severity of diseases related to these organisms may be high (botulism, cholera) or low (*Aeromonas* infections), but the likelihood of provoking diseases (risk) is extremely low. The pathogenic strains require temperatures $>1°$ C for growth and they are competing with the normal spoilage flora whose growth potential is comparatively much higher at low temperatures. Thus the products are likely to be spoiled before production of toxin or development of high numbers of pathogens. When the products are cooked before consumption, this will completely eliminate the risk.

Pathogenic bacteria from the animal/human reservoir (*Salmonella, E. coli, Shigella, Staphylococcus aureus*) may contaminate the live animal depending on the fishing area and further contamination may take place during landing and processing (Figure 5.2). The diseases which these organisms can provoke are serious, but if numbers on the products are low (i.e. no growth has taken place) the likelihood of this to happen (risk) is very low indeed. Cooking before consumption will eliminate the risk. However, an indirect hazard exists if contaminated products are polluting the working areas (industry, kitchen) and thereby transporting the pathogens to products which are *not* cooked before eating (cross contamination). This indirect hazard must also be prevented.

In contrast, the effect from growth of histamine producing bacteria (*Morganella morganii*) will not be eliminated by cooking or any other heat treatment as the heat resistance of histamine is high. The risk of histamine poisoning, if fish (*Scombroidae*) have been kept for some time at elevated temperatures ($>5°$ C) is therefore high.

Fish caught in certain areas may be infected with parasites dangerous to human health. The severity of the possible disease depends on the parasite involved, and the likelihood of contracting parasites from fish is eliminated if the fish is cooked before consumption. A low risk will exist if fish are consumed raw.

Figure 5.2. Exposure of catches to heavy contaminated coastal waters during landing

The presence of biotoxins and chemicals in fish depends on fish species, fishing area and season. The biotoxins are heat-stable and the risk of intoxication after consumption (raw or cooked) is high. The safety hazards related to fish raw materials for further processing and to consumption of fresh and frozen fish are summarized in Table 5.5.

Heavy contamination and in particular, growth of specific spoilage bacteria is certain to reduce the normal and expected shelf-life of the product (high risk). This may of course cause serious commercial problems but no lives are endangered. Thus, severity is low.

The Critical Control Points in production of fish and frozen fish are marked in Table 5.6.

Control of hazards and the environment

The contamination of live fish with bacteria normally found in the environment obviously cannot, and need not be controlled (- it is a hazard, but with no risk). However, contamination with bacteria from the animal/human reservoir can be limited by monitoring the fishing areas and control of fishing if gross pollution from population centres or industry is evident. More important, however, is the monitoring of the fishing areas for the presence of parasites, biotoxins (-toxic fish or toxic marine plankton), and toxic chemicals.

Table 5.5. Hazard analysis of fish raw materials and the processing of fresh and frozen fish products.

Organism/component of concern	Hazard			
	Contamination	Growth	Severity	Risk
Pathogenic bacteria				
indigenous	-	+	high/low	no risk[1]
non-indigenous	(+)	+	high	low
Virus	(+)	-	-	no risk[1] low
Biotoxins	+	-	high	high
Biogenic amines	-	+	low	high
Parasites	+	-	low	no risk [2]
Chemicals	+	-	low	low
Spoilage bacteria	(+)	+	low	high

1) no risk if product is cooked. 2) no risk if product is cooked or frozen.

Monitoring of the aquatic environment for pollution (fecal) and presence of chemical toxins and biotoxins in fish or algae may in most countries be a responsibility of the government and is most conveniently carried out by specialized laboratories. However, even with the best monitoring of the environment, the risk of toxic fish reaching the consumer can be reduced, but not completely eliminated. Thus, for this particular hazard, only a CCP-2 can be established. The critical limits for biotoxins and pollution are found in national legislations or international recommendations. The most important are quoted in Section 3.

Temperature control

Time and temperature (Txt) conditions at all times (all steps) from catching to distribution is a CCP-1 in preventing growth of pathogenic bacteria, histamine producing bacteria and spoilage bacteria. At $t < 1^\circ C$, no growth of pathogenic bacteria takes place. Only small and insignificant amounts of histamine may be formed and bacterial spoilage is not inhibited, but taking place at a "normal" and expected rate. A maximum time at $t > 5^\circ C$ (or max. processing time) must be specified in the criteria or tolerances for this CCP.

Time and temperature conditions are also important CCPs in preventing oxidation and chemical spoilage. Thus exposure for a few hours of fatty fish to sun, air and ambient temperature during e.g. catch handling, is sufficient to introduce severe quality loss and cause early chemical spoilage (Figure 5.3).

Table 5.6. Hazards and Critical Control Points in the production of fresh and frozen fish.

Product flow	Hazard	Preventive measure	Degree of control
Live fish	Contaminated[1)]	Monitoring of environment	CCP-2
Catch and catch handling	Growth of bacteria	(Txt) control	CCP-1
Chilling	Growth of bacteria	(Txt) control	CCP-1
Landing	Excess contamination and/or	Hygienic handling	CP
	growth of bacteria	(Txt) control	CCP-1
Arrival of raw material at factory	Substandard quality entering processing	Ensure reliable source	CCP-1
		Sensory evaluation	CCP-2
Storage of raw material			
Washing			
Filleting	Presence of parasites	Candling	CCP-2
Skinning			
All processing steps	Growth of bacteria	(Txt) control	CCP-1
	Contamination	Factory hygiene	CP
		Water quality	CCP-1
		Sanitation	CP
Packaging	Spoilage (oxidation)	Packaging material/vacuum	CCP-1
Chilling	Growth of bacteria	(Txt) control	CCP-1
Freezing	Chemical/autolytic spoilage	(Txt) control	CCP-2

1) The hazards are excess contamination with pathogenic bacteria (Group 2), biotoxins, parasites and chemicals.

Figure 5.3. Delay in icing chilling on board may cause bacterial growth (histamine formation, spoilage) and chemical spoilage (oxidation).

Monitoring of time/temperature conditions during handling and processing can be done by date marking of boxes and containers and by visual inspection of icing and chilling conditions. Time and temperature recording at specific points and during processing should preferably be controlled automatically. Process flow must be designed to avoid stops and interruptions, and all chill rooms must be supplied with thermometers. Visual inspection (e.g. quantity of ice) and control checks of temperature must be done in a daily routine. Automated time/temperature integrators are also available on the market and can be usefully applied. A log on temperature recordings (manually or automatically read) must be kept and be available at all times.

A sensory assessment (appearance, odour) of the raw material on reception at factory or immediately before processing is a CCP-2 for ensuring that until this point the material has been under control, that spoiled fish or shrimp do not enter the processing area, and that potential toxic species can be discarded.

Factory hygiene and sanitation

Adherence to initially established GMP is as well as sanitation and factory hygiene procedures are control points (CP) to reduce or avoid gross contamination and these control measures must be monitored as a daily routine (see Section 5.1.3.A). It should be noted that contamination during processing of raw fish to be eaten as cooked is a hazard with a very low or no risk (Table 5.5). Consequently, hygiene and sanitation in this type of production is not truly a CCP but only a control point (see Section 5.1.1.B for discriminating between CCP and CP).

Packaging and freezing are CCPs for control of chemical and autolytic spoilage. The packaging methods and materials (which are the criteria for this CCP) are normally specified in the sales contract. The freezing method is restricted to available equipment but rapid freezing to t $\leq$ -18°C and a storage temperature at -18°C are essential criteria for the second CCP.

All observations and measurements must be recorded on checklists and data sheets. Examples are shown in Tables 5.7 and 5.8 (after Hudak-Roos and Garrett 1992).

In conclusion, it can be stated that in the production of fresh and frozen fish and crustaceans most hazards can be controlled in a routine quality assurance programme using very simple equipment and methods. Only the presence of heat stable biotoxins remains a partly uncontrolled hazard.

Table 5.7. Examples of checklist for observations on sanitation (after Hudak-Roos and Garrett 1992).

CHECKLIST EXAMPLE

SANITATION LOG

S = Satisfactory
N = Needs Improvement
A = Alert

Date:_______

TIME:	Pre-Start	Break 1	Break 2	Comments
Thaw tank cleaned				
Glaze water changed				
Belts cleaned and in good repair				
Utensils cleaned and in good repair				
Processing machines cleaned				
Lighting				
Floor clean				
Ceilings without peeling paint or condensates				
Dip stations				
Trash removed				
Chlorine barrels				

Ins. By:_______ Prod. Supv.:_______ QA Mgr.:_______

Table 5.8 Example of data sheet on temperature (after Hudak-Roos and Garrett 1992)

DATA SHEET

I.T. LINE 1 C.L. = 180°F

TIME	TEMPERATURE
0800	181
0830	181
0900	180
0930	180
1000	179
1030	179

NOTES:

OPERATOR DATE

C. Lightly preserved fish products

This group includes fish products with low salt content (<6% NaCl (w/w) in aqueous phase) and low acid content (pH > 5.0). Other preservatives (sorbate, benzoate, NO_2, smoke) may or may not be added. The products may be prepared from raw or cooked raw material, but are normally consumed without prior heating. Product examples are salted, marinated and cold smoked fish. These products have a limited shelf life, even at chill storage - and there is ample epidemiological evidence of foodborne diseases being traced to these types of products. Nearly all known bacterial pathogens as well as production of biogenic amines are of concern. Parasites may survive and preformed bacterial toxins as well as biotoxins may remain stable during processing and storage of these products. The hazards related to this type of products are summarized in Table 5.9.

The contamination of lightly preserved fish products with **low** numbers of potentially pathogenic organisms normally found in the environment may or may not be regarded as a hazard. These organisms are frequently or always found on raw materials used and are as such very difficult or impossible to keep completely away from the final product. It must be emphasized, however, that these pathogens will contaminate the environment in the fish factory and that high numbers then may be found in any niche where the prevailing conditions (temperature, nutrients, etc.) are favourable for growth. Gross contamination of the final products may take place from these niches and the presence of **high** numbers of these organisms on products to be eaten without cooking is a hazard with a high risk which need a CCP (see also the discussion on control of *Listeria* in Section 3.1).

Table 5.9. Analysis of safety hazards in production of lightly preserved fish processing.

Organism/component of concern	Hazard			
	Contamination	Growth	Severity	Risk
Pathogenic bacteria				
indigenous	(+)	+	high/low	high
non-indigenous	+	+	high	high
Virus	+	-	high/low	high
Biotoxins	+	-	high	high
Biogenic amines	-	+	low	high
Parasites	+	-	low	high
Chemicals	+	-	high/low	low
Spoilage organisms	(+)	+	low	high

The CCPs for keeping the contamination of the final product with pathogenic microorganisms including virus at a low level is good factory hygiene. Monitoring of the environment, including the factory environment for these organisms, should be carried out at regular intervals depending on the local situation. GMP and factory hygiene must be specified in detail and monitored routinely (see also Section 5.1.3.A).

In contrast, any growth of pathogenic organisms including producers of biogenic amines, is a hazard of potential high severity and high risk. Thus this hazard must be controlled at all cost and production, distribution and storage are extremely important CCPs where time-temperature conditions must be controlled. For most pathogenic bacteria, the traditional chill storage at +5°C or lower is a CCP-1 but it should be emphasized that some of the pathogens are psychrotrophic in nature. This includes *L. monocytogenes* and *C. botulinum* type E which can multiply and produce toxins at temperatures down to +3.2°C. For the latter organism, an additional CCP is recommended. A salt level of at least 3% NaCl (w/w in aqueous phase) should be included in the criteria for production of lightly preserved fish products as this is sufficient to prevent growth and toxin production (Cann and Taylor 1979) at low temperature. The slightly increased risk caused by vacuum-packaging or storage of these products in oxygen free environment is insignificant if two separate CCPs of CCP-1 nature are applied and constantly monitored (temperature and salt content).

The presence of biotoxins and parasites in raw material for production of lightly preserved fish products is clearly a hazard of high or low risk depending on the fishing area and season. No CCP-1 can be identified for these hazards, but the risk from biotoxins can be reduced if the fishing grounds are monitored for the presence of toxic organisms (CCP-2) as discussed in the paragraph on fish raw material. As far as parasites are concerned, a "processing for safety" step (e.g. freezing of raw material) should be included in the process.

Spoilage is prevented by controlling raw material, time and temperature (Txt) conditions during processing and distribution, packaging material (oxygen transmission rate of film) and-method (degree of vacuum).

Table 5.10 summarizes the hazards and preventive measures during processing of cold smoked fish.

Table 5.10. Hazards and preventive measures in production of cold smoked salmon.

Product flow	Hazard	Preventive measure	Degree of control
Raw material before entering the factory	See Table 5.6	See Table 5.6	See Table 5.6
Reception of raw materials	Substandard quality entering processing	Ensure reliable source	CCP-2
Washing			
Filleting			
Salting	Salt content too high or too low (i.e. unacceptable taste or risk of growth and toxin production by *C. botulinum* respectively)	Visual observations of salting procedures and equipment Measurement of salt content of brine and product	CCP-2
Smoking			
Packaging	Spoilage (oxidation, microbial spoilage)	Visual control of packaging material and method (vacuum)	CCP-1
All processing steps	Growth of bacteria Contamination	(Txt) control Factory hygiene Water quality Sanitation	CCP-1 CCP-2 CCP-1 CCP-2
Chilling	Growth of bacteria	(Txt) control	CCP-1
Distribution	Growth of bacteria	(Txt) control	CCP-1

Note: The possible presence of live parasites is not controlled by these control measures. As there is no CCP-1 for this hazard in the normal production process, a period of freezing (-20°C for 24 hours) either of the raw material or of the final product must be included in the processing.

D. Heat treated (pasteurized) fish and shellfish products

A number of fish products receive a heat treatment during processing. Examples are: pasteurized or cooked and breaded fish fillets, cooked shrimp and crabmeat, cook-chill products and hot smoked fish. After the heat-treatment the various products may pass through further processing steps before being packed and stored/distributed as chilled or frozen products. Some of these products may receive additional heat treatment before consumption (cooked and breaded fillets, cook-chill products) or they may be eaten without further heat-treatment (hot smoked fish, cooked shrimp). Thus it is clear that some of these products are in the high risk category being extremely sensitive to contamination after the heat treatment.

To further illustrate the safety aspects, there is ample epidemiological evidence that this type of product has been the cause of food poisoning due to growth of coagulase-positive *Staphylococcus aureus* and enteropathogenic organisms among the *Enterobacteriaceae* and *Vibrionaceae*. Marine crustaceans, usually shrimp, crab or dishes made from them, accounted for 25 outbreaks of food-borne diseases reported in the United States during the period 1977-84 (Bryan 1988). Although there has been no recorded case of botulism caused by consumption of cooked shrimp, this possibility should not be overlooked, particularly in view of the variation in the ultimate use of this product.

In the application of the HACCP-system to these types of products, the heat-treatment is a very critical processing step. Hazards identified before this step may or may not be eliminated depending on the degree of heat being applied. Most criteria for heat-treatments have been laid down as a consequence of economical and technological considerations and not for hygienic or public health reasons. Noteworthy exceptions are the U.S.-regulations cited by Pace and Krumbiegel (1973) requiring a heat treatment of 82.2°C for a minimum of 30 min. in the processing of smoked fish with the purpose of killing all *C. botulinum* type E spores and the German requirement of the coldest part of the fish to be heated to 70°C to kill any nematodes, particularly *Anisakis simplex* present in the fish (German Fish Ordinance 1988). (Note: heating to 70°C is a gross overkill as 55°C for 1 min. is sufficient to kill nematode larvae - see Section 3.4).

Whenever possible, the heat-treatment should be utilized to eliminate harmful organisms. The criteria (time/temperature requirements) should be based on research demonstrating the lethal effect of the proposed heat-treatment.

Following this principle, any bacterial contamination and growth taking place **after** the heat-treatment is a serious hazard with a very high risk as shown in Table 5.11. In contrast, there are no hazards related to presence of parasites as there is no risk of recontamination with these components after the heat-treatment. A possible hazard related to presence of biotoxins and chemicals is dealt with under "fish raw materials for further processing" (see Section 5.1.3.B).

The critical control points during processing of heat treated products are therefore:

- The heat-treatment - which is a CCP-1 in control of pathogenic bacteria.

- GMP and factory hygiene/sanitation, which are CCP-2 in the control of recontamination and possible growth of bacteria **after** the heat treatment

- The water quality - which is a CCP-1 in avoiding any contamination from this source.

As an example, the CCPs in the processing of cooked, peeled and frozen shrimp are shown in Table 5.12.

Table 5.11. Hazard analysis in processing of heat treated (pasteurized) fish and shellfish products.

Organism/component of concern	Hazard			
	Contamination	Growth	Severity	Risk
Pathogenic bacteria				
indigenous	+	+	high/low	high/no risk[1)]
non-indigenous	+	+	high	high/no risk[1)]
Virus	+	-	high	high/no risk[1)]
Biotoxins	+	-	high	high
Biogenic amines	-	+	low	high
Parasites	+	-	low	no risk
Chemicals	+	-	high/low	low

1) there is no risk if products are cooked again immediately before consumption.

The criteria and the critical limits to be used in monitoring the CCPs are important and must be specified in detail. Thus the heating conditions (water temperature, belt speed etc.) necessary to obtain the desired result (e.g. minimal internal temperature of 80°C for 2 min.) must be determined by experimentation. Similarly the requirements to GMP, factory hygiene and sanitation procedures must be determined and described in detail as criteria for these CCPs. Once determined and precisely described, the daily monitoring can easily be carried out by visual observations and occasionally microbiological testing of cleaned surfaces (see also Section 5.1.3.A).

Table 5.12. Hazards and preventive measures in production of cooked, peeled, individually quick frozen (IQF) shrimp.

Product flow	Hazard	Preventive measure	Degree of control
Live shrimp			
Catch and catch handling	Black spots / Excess chemical preservative (sulfite)	Correct treatment with preservative (sulfite)	CCP-2
Chilling	Growth of bacteria	(Txt) control	CCP-1
Transport	Growth of bacteria	(Txt) control	CCP-1
Reception of raw materials	Substandard quality entering processing	Ensure reliable source, sorting	CCP-2
Washing			
Sorting			
Cooking	Over-/undercooking (i.e. loss of yield and quality - survival of bacteria)	(Txt) control	CCP-1
Peeling			
Separation/ cleaning			
Freezing (IQF)			
Packaging			
All processing steps after cooking	Recontamination	Factory hygiene Water quality Sanitation	CCP-2 CCP-1 CCP-2
Frozen storage	Loss of quality	Temperature control	CCP-2

E. Heat processed (sterilized) fish products packed in sealed containers (canned fish)

The basis for canning is the use of thermal processing to achieve commercial sterility of the final product. The containers are distributed at ambient temperature and often stored for months, even years under these conditions. The content of the cans are normally eaten without any heating immediately before consumption. Thus the hazards related to this products are:

- Survival of pathogens during heat processing.

- Presence of heat-stable toxins (biotoxins, histamine) in the raw material.

- Recontamination of product after heat processing (faulty containers, poor sealing, contaminated cooling-water, faulty container handling).

The Critical Control Points during production of canned fish are shown in Table 5.13.

The incoming raw material may be contaminated with biotoxins, histamine or toxic chemicals. There is no CCP-1 for these hazards during processing and proper control is therefore necessary at an earlier phase as described under Section 5.1.3.B. Similarly, the quality of the cans should be ensured by a documented quality assurance system by the can manufacturer. Additionally visual observations can be made. Guidance on visual inspection of cans will be found in Fisheries and Oceans (1983), AOAC/FDA (1984) and Thorpe and Barker (1984).

Correct filling is important to ensure proper heat penetration, hence it is a CCP.

A hermetically sealed container is a prime requirement, and control of this operation is paramount. Many types of closing machines are in use and it is essential that each machine is operating properly and trained mechanics keep them under control. Standards of can closure must be checked at regular intervals and always when setting up a new machine or following adjustment of an old one. It is normally recommended for metal cans, that tear down measurements be made once per shift and a formal/visual examination every half hour (ICMSF 1988, Varnam and Evans 1991). Details of seam examination may be obtained from Anon. (1973) and Hersom and Hulland (1980).

The thermal processing is a CCP-1 for eliminating all pathogenic organisms. Most processes are set to destroy spores of *C. botulinum* and are based on the so-called "botulinum cook" (F_o = 3, see Section 3.1). Monitoring of this CCP can be considered in 2 phases. The first concerns the pre-processing operations such as control of product temperature before retorting, control of time between container closure and retorting, loading of the retort, attachment of heat sensitive tape, venting of retort. The second phase is the thermal process.

This includes control of operational requirements such as steam pressure, water circulation and chain speed. The thermal processing should be timed from 2 points: the start of the heating and the point at which the sterilization temperature is reached. Properly calibrated thermometers must be used (platinum resistance thermometers are increasingly used as being the most accurate).

Table 5.13. Hazards and preventive measures in production of low-acid canned fish.

Product flow	Hazard	Preventive measure	Degree of control
Raw material before entering the factory	See Table 5.6	See Table 5.6	See Table 5.6
Reception of raw material at factory (fish and cans)	Substandard quality entering processing	Ensure reliable source Sensory evaluation	CCP-2
Primary processing			
Filling of cans	Uncontrolled heat penetration during thermal processing	Avoid inclusion of air, control weights of solids, liquids, product density and headspace	CCP-2
Evacuation, seaming	Recontamination	Standards of closures must be checked at regular intervals	CCP-2
Thermal processing	Survival of pathogens	(Txt) control	CCP-1
Cooling	Recontamination	Quality of cooling water chlorine level >1-2 ppm	CCP-2
Handling of filled (wet) cans	Recontamination	Handling of warm, wet cans must be avoided Can handling should be designed to minimize mechanical shock	CCP-2
Storage and distribution			

The ICMSF (1988) has summarized the monitoring requirements:

- Date, code, product, container size, no. of cans in retort.
- Time venting starts and ends.
- Time sterilization temperature starts and ends.
- Sterilization temperature (read from master thermometer).
- Pressure at sterilization temperature.
- Time steam off.
- Elapsed time between steam on and off.
- Time retort is opened.
- Container status indicator check.
- Name of retort operator.
- Cross reference to recorder chart.

For other types of retorts, special problems may exist see FAO/WHO International code of practice for low-acid and acidified canned food (FAO/WHO 1979).

The cooling operation is a CCP-2 for preventing contamination by the cooling medium. A high standard of hygiene must be maintained and cooling water must be chlorinated. Before water is used for cooling, it should have at least 20 min. contact with 1-2 ppm. of free chlorine.

Residual chlorine measurements must also be made **after** cooling water has been in contact with cans. Additionally, microbiological analysis of the cooling water may be carried out. The mesophilic aerobic plate count must be less than 100 CFU/ml (ICMSF 1988).

The warm damp cans may readily be infected if exposed to excessive contamination in the area of the seams. Container handling is therefore a CCP-2. Handling of warm, damp cans should be avoided and possible contact surfaces must be thoroughly cleaned. Excessive physical handling of cans must also be avoided.

There are no hazards related to storage and distribution of final product. However, it is common practice - and in some cases a legal requirement (e.g. EEC Directive 91/493/EEC (EEC 1991b)) that checks must be carried out at random by the manufacturers to ensure that products have undergone appropriate heat treatment. This requirement is conveniently included as part of the verification procedures and involves taking random samples of the final product for:

- Incubation tests. Incubation must be carried out at 37°C for seven days or at 35°C for ten days or at any other equivalent combination.

- Microbiological examination of contents and containers in establishments laboratory or in another approved laboratory.

F. Semi-preserved fish

These are fish products with a salt content >6% NaCl (w/w) in the aqueous phase or a pH <5.0. Preservatives (sorbate, benzoate, nitrate) may or may not be added. These products still require chill storage and may have a shelf life of 6 months or more. Normally there is no heat-treatment applied neither during processing nor in the preparation before consumption. Traditional production often includes a long ripening period (several months) of the raw material before final processing. Product examples are salted and marinated fish, fermented fish and caviar products.

The contamination of these products with pathogenic bacteria is not a hazard. Also growth of these organisms is completely inhibited if storage temperature is kept <10°C. The mesophilic proteolytic *C. botulinum* (type A and B) and *Staphylococcus aureus* which are otherwise able to grow at the high NaCl-concentration in these products, cannot grow at temperatures below 10°C.

Nevertheless there is epidemiological evidence that these products have been the cause of a number of foodborne diseases related to the presence of biotoxins including histamine, bacterial toxins and parasites.

C. botulinum toxins are stable at high salt and low pH (Huss and Rye Petersen 1980). Thus any toxin present or preformed in the raw material will be carried over to the final product. These hazards can only be controlled by having full control over the raw material handling as described under Section 5.1.3.B. If this is not possible, the raw materials for this production is a CCP-2, but methods for monitoring are very limited. A sensory evaluation will give some indication - but no guarantee that no toxin (histamine, botulinum toxin) is present.

In contrast, the presence of live parasites in these products is a hazard which can easily be controlled. The requirements for salt level and holding times are presented in Section 3.4 and if these cannot be met, a freezing step must be included as mentioned for lightly preserved fish (this Section). An example of CCPs in processing of semipreserved fish is shown in Table 5.14.

G. Dried, dry-salted, smoke-dried fish

These are products with very high salt content (saturated NaCl in water phase) and/or very low water activity due to drying. The products are stable at ambient temperature and may be eaten after rehydration and cooking or directly without any cooking. The hazard related to processing is primarily a time factor. Processing takes place at ambient temperature, and if lowering of water activity takes too long, growth and toxin production by microorganisms will take place.

However, hazards related to the raw materials (production of bacterial toxins and histamine, presence of biotoxins and chemical toxins) may also be carried over to the final products. These hazards must be controlled as described for "raw materials for further processing".

Salted or dried fish may spoil due to growth of halophilic bacteria ("pink") or moulds ("dun"). These microorganisms may be introduced with the salt or via contamination from poorly cleaned equipment and utensils used during processing. Prevention measures (CCP-2s) are good factory hygiene and sanitation, and if possible, storage at < 10°C, which is a CCP-1 for this hazard.

Table 5.14. Hazards and preventive measures in production of marinated herring.

Product flow	Hazard	Preventive measure	Degree of control
Raw materials before entering the factory	See Table 5.6	See Table 5.6	See Table 5.6
Reception of raw material at factory	Substandard quality entering processing	Ensure reliable source Sensory evaluation	CCP-2
Primary processing			
Filleting			
Salting in brine	Incorrect salt content in fish (spoilage and/or survival of parasites)	Control of salt concentration in brine and time for fish in brine (NaCl concentration and holding time to be specified).	CCP-1
Marinating	Incorrect NaCl and acetic acid concentration in fish (taste, spoilage and/or survival of parasites)	Control of composition of marinade and marinating time. Holding time to be specified	CCP-1
Secondary processing			
Packaging in glass jars in final pickle	Poor sensory quality	Control of composition of pickle (concentration of sugar, acetic acid, spices, etc.)	CCP-1
Distribution	Growth of microorganisms (bacteria, yeasts) (spoilage, toxin production by *C. botulinum* type A, B).	(Control of temperature T < 10°C)	CCP-1

5.1.4. Fish regulations, regulatory agencies and HACCP

The HACCP-concept has been applied very successfully by the U.S. Food and Drug Administration (FDA) since 1973 in the control of microbiological hazards in low acid canned food (FDA 1973). No other regulatory agency considered including HACCP in their food safety programs until it was strongly recommended by a subcommittee on microbiological criteria set up by the U.S. National Research Council (FNB/NRC 1985). Following that, the U.S. National Marine Fisheries Service investigated the mandatory use of HACCP in the seafood industry in a study entitled the "Model Seafood Surveillance Project" (Garrett and Hudak Roos 1991). Also in Canada, a new Quality Management System which embodies the HACCP philosophy, has been introduced and mandatory since February 1993 (White and Noseworthy 1992).

The principles of HACCP can easily be incorporated into national fish regulations, but it should be emphasized that HACCP deals with the uniqueness, while regulatory agencies are used to deal with the general issues which can be formulated in regulations to cover the whole industry. A HACCP system needs to be tailored to each individual plant and each processing line. This calls for close cooperation between regulatory agencies and food industry, which is not easy to achieve. Highly educated people and staff trained in the application of HACCP are needed, as well as mutual respect, understanding and trust - on both sides.

Once established, each particular plant needs to have the system approved by the regulatory authority having jurisdiction. All CCPs and monitoring records can then be checked by inspectors and compliance with safe processing requirements can easily be confirmed. As a further guarantee, the regulatory agencies may also occasionally carry out verification tests to ensure that the HACCP system is working. The U.S. National Advisory Committee on Microbiological Criteria for Foods (NACMCF 1992) has stated that governments' regulatory responsibility is part of the verification activities and give the example shown below:

Examples of verification activities

A. Verification procedures may include:

- Establishment of appropriate verification inspection schedules.
- Review of the HACCP plan.
- Review of CCP records.
- Review of deviations and dispositions.
- Visual inspections of operations to observe if CCPs are under control.
- Random sample collection and analysis.
- Review of critical limits to verify that they are adequate to control hazards.
- Review written records of verification inspections which certifies compliance with the HACCP plan or deviations from the plan and the corrective actions taken.
- Validation of HACCP plan, including on-site review and verification of flow diagrams and CCPs.
- Review of modifications of the HACCP plan.

B. Verification inspections should be conducted:

- Routinely, or on an unannounced basis, to assure selected CCPs are under control.
- When it is determined that intensive coverage of a specific commodity is needed because of new information concerning food safety.
- When foods produced have been implicated as a vehicle of foodborne disease.
- When requested on a consultative basis or established criteria have not been met.
- To verify that changes have been implemented correctly after a HACCP plan has been modified.

C. Verification reports should include information about:

- Existence of a HACCP plan and the person(s) responsible for administering and updating the HACCP plan.
- The status of records associated with CCP monitoring.
- Direct monitoring data of the CCP while in operation.
- Certification that monitoring equipment is properly calibrated and in working order.
- Deviations and corrective actions.
- Any samples analyzed to verify that CCPs are under control. Analyses may involve physical, chemical, microbiological or organoleptic methods.
- Modifications to the HACCP plan.
- Training and knowledge of individuals responsible for monitoring CCPs.

Regulatory agency and industry cooperation can provide industry and governments control systems as well as potential buyers of the products with the necessary **confidence** in the quality assurance programme. Once the confidence is established, the entrance of such products to the world market can be significantly simplified by means of the signing of M.O.U.s (Memorandum of Understanding) between countries and between importers and exporters. One further advantage is that duplicating control effort can be avoided resulting in cost/benefit value to both parties.

A particularly sensitive issue in this process is that regulatory agencies must have access to industry records. This point of major disagreement needs to be resolved. There can be no discussion that the regulators should have access to monitoring results of CCPs and actions taken, while some information related to manufacturing practices may be proprietary.

5.1.5. Advantages and problems in the use of HACCP

The great advantage of the HACCP-system is that it constitutes a systematic, structural, rational, multi-disciplined, adaptable and cost-effective approach of preventive quality assurance. Properly applied, there is no other system or method which can provide the same degree of safety and assurance of quality and the daily running cost of a HACCP system is small compared with a large sampling programme.

By using the HACCP-concept in food processing it is possible to assure and to document assurance of **minimum** quality standard such as:

- Absolutely safe food. If absolute safety cannot be guaranteed (e.g. consumption of raw molluscs), it is clearly disclosed by the programme, and a general warning should be given.

- Products having a stated and agreed (normal) shelflife, if handled and stored according to instructions.

Further advantages which are evident from the text (Mitchell 1992) have been listed and summarized:

1. Control is proactive in that remedial action can be taken before problems occur.

2. Control is by features that are easy to monitor, such as time, temperature and appearance.

3. Control is fast such that prompt remedial action can be taken if necessary.

4. Control is cheap in comparison with chemical and microbiological methods of analysis.

5. The operation is controlled by those persons directly involved with the food.

6. Many more measurements can be taken for each batch of product because control is focussed at the critical points in the operation.

7. HACCP can be used to predict potential hazards.

8. HACCP involves all levels of staff in product safety, including non-technical personnel.

The general principle of the HACCP-concept is to direct energy and resources towards areas where they are necessary and most useful (i.e. "distinguish between the nice and the necessary"). This idea makes HACCP an ideal tool where resources are scarce as is the case in many developing countries. It may appear an immense and impossible target to upgrade an underdeveloped industry to be able to produce safe food for export. However, by using the HACCP-concept it is possible to identify the **necessary** changes in procedures and/or new installations.

For small plants, processing fresh fish only, all that is needed may be absolute temperature control from catching/landing to distribution (see also Section 7.5).

However, as the HACCP concept was developed more than 20 years ago, it is a very relevant question to ask: Why is it that this system is not in general use all over the world? The following are some problems that need to be considered (Tompkin 1990):

- There continues to be a non-uniform understanding of HACCP both nationally and internationally. New definitions and principles are evolving as a result of extensive debate. At times it appears that principles are reversed to extensive sampling and "regulation" of every minute detail. Producer - buyer and producer - regulator disagreements over end-product testing will not disappear and difference of opinion will exist over the degree to which HACCP can replace the need for end-product testing.

- There is no universal agreement on what constitutes a hazard (e.g. presence of *Listeria monocytogenes* in raw food). There is an urgent need to establish an international body consisting of members being non-political, but highly respected scientifically, who can advise on safety issues and current scientific thinking on hazards in foods.

- To be effective, HACCP needs to be applied from origin of food (sea/farm) to consumption. This may not always be possible.

- HACCP deals with the uniqueness - regulations with the general. This problem may be difficult for regulatory agencies to understand and accept, thereby delaying the application of the system.

- Acceptance of HACCP requires mutual trust. Unless this trust exists or can be created between the regulator and the regulated, the system will fail.

- HACCP requires processors to accept greater responsibility. This may cause some resistance from processors who normally rely on government services (inspectors, laboratories) to guarantee for safety and quality. Furthermore, it can lead to a perception that HACCP results in reduced inspection and loss of regulatory control even though the intent of HACCP is just the opposite.

- It will take a long time to train inspectors and industry personnel to arrive at one common understanding of HACCP.

- Application of HACCP will not prevent all problems, and "experts" may disagree on vital issues.

The decisions and priorities on issues related to health hazards are influenced by a number of factors. The scientific community can only provide one dimension, and even requires that all sound scientific data can be uniformly interpreted. However, the risk perception and emotional feelings by the consumers are often quite different and the producers are by nature mostly concerned about cost and competing markets.

The legislative and regulatory authorities are the institutions who must sort the information and set the rules. This means that these agencies must be staffed with highly educated and trained people, who can keep abreast with the latest scientific development. These institutions must be completely independent from commercial and other interests which may affect their decisions and also try to avoid the bureaucratic pitfalls spending all their time and effort making regulations and controlling matters of minor importance such as tiles on the walls, types of watertaps to be used and the numbers of doors leading into a particular room. Such non-HACCP attitudes by regulatory agencies may be further aggravated in a democracy, where these agencies may overreact to minor health hazards about which there is a strong public concern, while taking little or no action in regard to health hazards that scientists have demonstrated to be of major importance, but arousing little public concern (Mossel and Drake

1990). A typical example is the large public concern and exaggerated regulatory response regarding authorized food additives, while this scientifically is identified as a minor problem.

In conclusion, it can be stated that for the HACCP concept to be truly operational and generally applied there is a great need for increased communication and understanding between the scientific community, the general public and the regulatory agencies. Only then can a better prevention of foodborne diseases be achieved.

5.2. APPLICATION OF THE ISO-9000 SERIES AND CERTIFICATION

This Section is prepared by Professor Mogens Jakobsen

5.2.1. Definition of ISO quality standards

The International Standards Organization (ISO) is located in Geneva, Switzerland, and is a federation of national standard bodies representing almost 100 countries.

Based on the good experience gained with the British Standards (BS) 5750 series published in 1979 they were adopted by ISO, and the ISO 9000 series were published in 1987 aiming at providing an international acknowledgement of quality efforts. Today, more than 50 countries have adopted the ISO 9000 series which is equivalent to the BS 5750 standards as just mentioned. In the United States, the standards are published as the ANSI/ASQC Q 90 series and in the European Community they are published as the European Norm (EN) 29000 Series.

The ISO 9000 series include 5 separated standards as shown in Table 5.15.

Table 5.15. ISO 9000 Series.

ISO Standard	Field of application
ISO 9000	Selection of the appropriate ISO 9000 standard
ISO 9001	Quality system requirements for product development, production, delivery and after sales functions
ISO 9002	Quality system requirements for production and delivery
ISO 9003	Quality system requirements for final inspection and testing
ISO 9004	Guidelines for ISO 9000, quality system elements

ISO 9001, 9002 and 9003 are three specific standards describing the elements and requirements of a quality system to be implemented in a company in connection with a contractual situation, i.e. the supplier - customer relation. They standardize and outline how companies can establish efficient Quality Systems, and they form the background for obtaining a Quality System Certificate issued by an approved, independent organization (certifying body).

As seen from Table 5.15, ISO 9001 is the most comprehensive standard containing most of the elements described in the guidelines given in ISO 9004. Compared to ISO 9002 the most important difference will be that it includes development of new products and processes. The ISO 9003 is used in situations where requirements to the producer only comprise final product inspection and testing and this standard only contains a minor part of the elements of ISO 9004.

For food processing companies, the most relevant standards will be ISO 9001 and 9002 containing the elements shown in Table 5.15 and described briefly in the following paragraphs.

However, it shall be mentioned that the combined use of standards can be advantageous. For small enterprises like a fishing boat, one could use ISO 9003 and add the relevant elements of ISO 9002. This can lead to the most appropriate system manageable by such a small entity. In such a case the official certification as mentioned above, will be according to the ISO 9003.

5.2.2. Elements of the quality system

The various elements of the ISO 9000 standards are shown in Table 5.16 and shall be commented briefly in the following.

Management responsibility is the first and overall most important system requirement mentioned. Full commitment of the company top management is a must, and the entire system must be under management control and review. Management shall define the objectives and the policy of the system, and it bears the full responsibility for ensuring that the policy is understood, implemented and maintained at all levels in the company. Responsibility and authority of all personnel who manage, perform and verify work affecting quality, shall be defined by the management, and adequate resources should be provided.

If the HACCP concept is applied, it shall be stated in the company quality objective and policy.

Requirement No. 2, which has the heading, **Quality System**, refers to the documented system ensuring that products comply with the specified requirements. It states that management shall ensure the presence of documented procedures and instructions in accordance with the ISO 9000 standard in question as well as efficient implementation of the Quality System procedures and instructions. As mentioned later and shown in Figure 5.4, the system will often be organized in three levels comprising the Quality Manual, Procedures and Instructions. If the HACCP concept is incorporated with the more narrow quality objective, e.g. with *Salmonella* as the defined risk, only procedures and instructions for control of *Salmonella* as defined in the objective of the system will be included.

In **Contract review** Requirement No. 3, it is laid down that the producer shall review and evaluate all contracts to ensure that he can supply a product which meets the customer's specified requirements and expectations, e.g. the product shall conform to specified requirements which for *Salmonella*, could be "absence in 25 g frozen shrimps" in each of a certain number of packages according to the sampling plan agreed upon.

Obviously, this element is very important in the ISO 9000 system designed to deal with supplier - customer relations. It is also stipulated that records of such contract reviews shall be maintained.

For **product development** Requirement No. 4, the supplier shall establish and maintain procedures which control and verify all phases of product development to ensure that the specified requirements are met. Using HACCP and *Salmonella* as an example, this means that the system shall ensure that new products and processes are not implemented unless they provide safety against *Salmonella* as laid down in the quality objective of the system. This is a complicated, demanding element of the standard, difficult to implement as well as maintain in the company. For the example used, it requires profound microbiological expertise.

Documentation is a vital part of the system, and so is **Document Control** as mentioned in Requirement No. 5. This control shall ensure that all necessary documents (procedures, instructions, forms, etc.) are available where needed, and obsolete documents promptly removed from all locations.

It is a key point of ISO 9000 that **Purchasing** (Requirement No. 6) only takes place from approved suppliers which have been selected on the basis of previous performance and an effective control system as well as their ability to meet the specified requirements. When applying the HACCP concept for the *Salmonella* in farmed frozen shrimp, it means that feed for the farm should only be purchased from feed mills producing feeds which do not contain *Salmonella* according to the specifications agreed upon. The standard reaches even further in demanding mutual cooperation and a contractual understanding with the feed mill; the mill shall be assessed to be included in the list of approved suppliers established according to ISO 9000 requirements. The feed mill shall be audited just like all other suppliers on the list, and purchased products shall be inspected on receipt, and feed-back on performance at all points shall be ensured. The very obvious reasons for these detailed requirements for purchasing obviously will be the inevitable effect of raw materials, machines, cleaning agents, services, etc. on the quality of the final product.

Procedures for product identification and traceability during all stages shall be established, maintained and recorded as stated in Requirement No. 7. If required, each batch, package, etc. shall be provided with a unique identification which shall be recorded.

Process Control (Requirement No. 8) shall ensure that all processes influencing the quality of the final product shall be specified and documented to ensure and verify that they are carried out under controlled conditions. This involves documented work instructions, including cleaning and disinfection procedures, use of appropriate equipment, machines, material and arrangement of processing facilities as well as monitoring of products and processes.

This element will be the key area for the HACCP concept with the Hazard Analysis, identification of Critical Control Points (CCPs) and monitoring of CCPs, as described in Section 5.1.

A schedule for **testing and inspection** of raw materials, intermediate and final products shall be established (Requirement No. 9). For the HACCP system, the schedule must be based on CCPs as identified in the hazard analysis. Test methods must be defined. Responsibilities for sampling and testing, reporting and control of non-conforming products shall be defined and reference made to the appropriate specifications.

The **test equipment** used shall be selected to demonstrate acceptable compliance with the defined specifications for the products and shall be calibrated at regular intervals against nationally recognized standard references (Requirement No. 10).

Proper identification of the **inspection and test status** shall be ensured with untested, tested, approved or rejected products being clearly marked (Requirement No. 11).

Procedures and instructions shall be established for **control of non-conforming products** (Requirement No. 12). In the present example, shrimp containing *Salmonella* will be a non-conforming product according to the specifications agreed upon. Such a product shall be identified, placed and labelled in a way that clearly isolates it and prevents it from being supplied as *Salmonella*-free by mistake. The responsibility for making decisions on disposition of non-conforming products shall be defined and documented. A non-conformity report shall be worked out stating the nature of the non-conformity, the disposition decided, and the corrective action to be initiated for resolving the non-conformity as described in the following.

The **corrective action system** (Requirement No. 13) is concerned with revising work operations etc. to try to eliminate the causes of failure. This is the system requirement that helps a company getting better and better by aiming at doing everything right the first time. To control all the activities required in the corrective action, forms containing the following points shall be applied: Clear statements of the non-conformity, responsibilities, action to be taken, date of implementation, verification and recording of the resulting new procedures.

Handling, storage, packaging and delivery (Requirement No. 14) is obviously very important for foods, to prevent damage or deterioration of the products. Temperature control including monitoring and recording shall be mentioned as examples to illustrate the importance of this requirement which obviously apply to all stages from raw materials, throughout production and delivery, and to the point of consumption. Determination and control of shelf life is needed, and so is full traceability with respect to the risk of product recall.

As mentioned several times in the above, recording is required with the purpose of demonstrating the achievement of the required quality and to demonstrate that the quality system is effective. This is stated in Requirement No. 15 **Quality Records** and the meaning of it will be seen from the following examples of records to be included: inspection reports, analytical results, calibration reports, audit reports and corrective action reports.

It is also demanded that the system be internally audited on a regular basis (Requirement No. 16, **Internal Quality Audits**). An appropriate audit plan shall be worked with ensuring that all elements (not necessarily all details) are audited e.g. once a year. Audit teams shall be formed and it must be assured the members are independent of the activities being audited. The audit report shall be included in the quality records as mentioned above.

Management shall carry out its own independent review and evaluation of the Quality System. This has to be carried out on a regular basis e.g. twice per year and it should be based on the above international audit reports mentioned above as well as evaluation of the overall effectiveness of the system in achieving the quality objectives stated. Needs for updating, new strategies etc. should also be indicated. It all has to be documented in a report. This again demonstrates the very active role to be played by the management of the company.

Training (Requirement No. 17) is a vital part of the ISO 9000 standards. Of similar importance to food companies are cleaning and disinfection and personal hygiene. These subjects have been included as separate requirements (No. 18 and 19) in Table 5.15 to emphasize their importance and they have been used for the following illustration of the structure of the system with its various types of documents.

Table 5.16. Quality system elements.

Quality System Requirements		Contents
1	Management responsibility	Define and document commitment, policy and objectives, responsibility and authority, verification resources and personnel. Appoint a management representative and conduct regular reviews of the system
2	Quality system	Establish and maintain a documented quality system ensuring that products conform to specified requirements
3	Contract Review	Ensure that customer's contractual requirements are evaluated and met
4	Product development	Plan, control and verify product development to ensure that specified requirements are met
5	Document control	System for control and identification of all documents regarding quality, e.g. procedures, instructions, and specifications
6	Purchasing	Ensure that purchased products conform to specified requirements

(cont.)

Table 5.16. Quality system elements (cont.).

Quality System Requirements		Contents
7	Product identification and traceability	System to identify and control traceability of product at all stages from raw materials through production to the final product as delivered to the customer
8	Process control	Ensure and plan the control of production which directly effects quality by documented work instructions, monitoring and control of processes
9	Inspection and testing	Inspect and test incoming products, intermediate and final product; establish product conformance to specified requirements and identify non-conforming products; maintain inspection and test records
10	Inspection, measuring and test equipment	Selection and control of equipment to ensure reliability and accuracy in measuring data
11	Inspection and test status	For the whole process the products shall be identified and clearly marked concerning test status, including indication of conformance or non-conformance
12	Control of non-conforming products	Identification, documentation, evaluation, isolation (if possible) and disposition of non-conforming products
13	Corrective actions	Prevention of reoccurrence of failures (non-conformance)
14	Handling, storage, packaging and delivery	Protection of the quality of the product during handling, storage, packaging and delivery
15	Quality records	Records, including those which demonstrate that the specified requirements have been met, shall be controlled and maintained
16	Internal Quality Audits	Regular, planned internal audits shall be carried out, documented and recorded to verify the effectiveness of the quality system
17	Training	Training requirements at all levels shall be identified and the training planned, conducted and recorded
18	Cleaning and Disinfection	Although not required by the ISO 9000 standards, these two points should be given special attention in all food companies
19	Personal hygiene	

5.2.3 The documented quality system

As mentioned above (Table 5.16, Requirement No. 2) the ISO 9000 standards require a documented quality system.

The three-level structure of documentation shown in Figure 5.4 has proved effective in the food industry as well as in other industries.

Level 1 is described in the Quality Manual. It is normally a short easy-to-read manual briefly stating the company's quality objectives and policies. All requirements of the appropriate ISO Standard will be addressed. The Quality Manual need not contain confidential information, and it is intended to be handed out to potential customers and others, to inspire confidence in the company being able to satisfy customers' expectations. In the example chosen, a proposal for Chapter 18, Cleaning and Disinfection, is shown in Table 5.17 for the plant producing frozen shrimps. The Table also shows the various formal requirements for the documents of the Quality System. Normally, the individual pages of the Quality Manual will be signed by the managing director or by the chairman of the board to demonstrate the commitment of the top management.

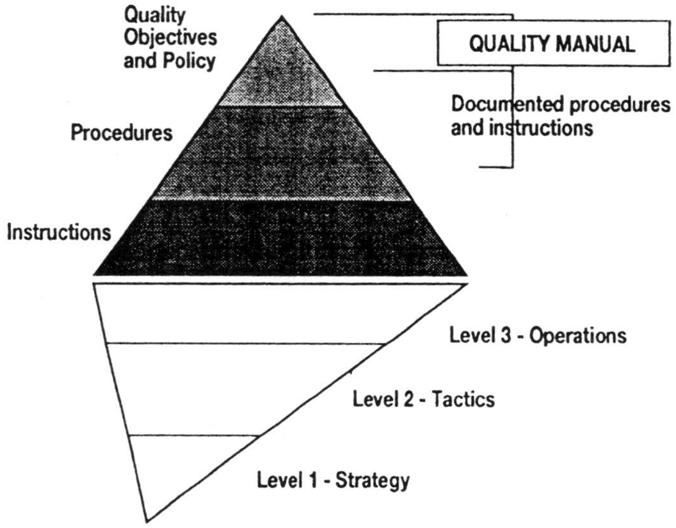

Figure 5.4. Typical structure of the Quality System.

The second level will comprise procedures, describing how the statements of the Quality Manual are deployed and implemented in the company. Persons responsible, as well as **where** and **when** will be stated. An example of a procedure is shown in Table 5.18. The above mentioned Chapter 18 of the Quality Manual underlies this procedure which could be issued by the QC-Manager and approved by the Technical Director.

The third level will comprise the work instructions giving all details on **how** the contents of the procedures are accomplished. Table 5.19 shows an instruction linked with the procedure featured in Table 5.18.

In levels 2 and 3, appropriate references will be given to various forms to be filled out e.g. the list of approved suppliers mentioned earlier and being part of the system documentation. An overview of the categories of documents included in a Quality System is given in Table 5.20. The Table almost speaks for itself and it clearly underlines the requirements for documentation including record keeping.

Table 5.17. Example of company policy (cleaning and disinfection). Quality manual, Chapter 18, level 1. Figure 5.4.

		CLEANING AND DISINFECTION
QUALITY MANUAL CHAP. 18	*REVISION NO:* 3	*DATE:* **1993-03-16**
EDITION: 1 **1993-01-15**		*PAGE:* 1 *of* 1

CLEANING AND DISINFECTION

It is the policy of Company XX to maintain a high standard of hygiene.

Procedures and instructions will be maintained to ensure that the high standard is in accordance with the requirements specified.

Detergents and disinfectants shall be selected and disinfection procedures worked out to ensure that the plant is free from Salmonella after cleaning and disinfection

ISSUED BY:	***APPROVED BY:***

Table 5.18. Example of Procedure, Cleaning and Disinfection. Level 2, Figure 5.4.

CLEANING AND DISINFECTION

PROCEDURE NO.: P 18 10 05 ***REVISION NO.:*** **3** ***DATE:*** **1993-03-16**

1. EDITION: 1993-01-15 ***PAGE:*** 1 ***of*** 1

CLEANING AND DISINFECTION

1.0 OBJECTIVE

Description of the procedure for cleaning and disinfection to ensure that the processing plant is visibly clean and Salmonella non-detectable after cleaning and disinfection.

2.0 RESPONSIBILITIES

The Technical Director is responsible for implementing and maintaining this procedure.

3.0 AREA OF APPLICATION

This procedure is valid for all areas, equipment, etc. where shrimp are handled within Company XX.

4.0 CLEANING AND DISINFECTION

With reference to the Technical Director, the foreman of the various sections of the processing plant are responsible for cleaning and disinfection.

Cleaning and disinfection is carried out at the end of each working day. The QC-Manager shall select and organize the use of cleaning agents and disinfectants in order to eliminate Salmonella, to prevent the build-up of scales or other residues as well as resistant microbial populations.

The QC-Manager is responsible for control and monitoring of the effectiveness of the cleaning and disinfection performed.

5.0 REPORTING

The QC-Manager and the foreman shall report their findings to the Technical Director

ISSUED BY: ***APPROVED BY:***

Table 5.19. Example of work instruction for cleaning and disinfection. Level 3, Figure 5.4.

CLEANING AND DISINFECTION

INSTRUCTION NO.: P 18 10 05 ***REVISION NO.:*** **3** ***DATE:*** **1993-03-16**

1. EDITION: 1993-01-15 ***PAGE:*** 1 ***of*** 2

CLEANING AND DISINFECTION

CLEANING AND DISINFECTION OF COOLER AND CONVEYER FOR COOKED SHRIMP

1.0 OBJECTIVE

It is the objective of this instruction to describe the cleaning and disinfection of cooler and conveyer for cooked shrimp carried out at the end of each working day, and before production in case the plant has not been in operation for more than two days.

2.0 RESPONSIBILITY

The Technical Director is responsible for implementing and maintaining this instruction.

The Foreman of Processing Hall C is responsible for carrying out this instruction.

3.0 AREA OF APPLICATION

This instruction is valid for Processing Hall C.

4.0 WORK DESCRIPTION

1.	Preparations	Cooler and conveyer are emptied and dismantled to allow cleaning of all parts.
2.	Rinse	Hosing with cold water
3.	Cleaning	Alkaline detergent "ZZ" is applied to all surfaces. Dosage: 3 l in 50 l cold water. pH: 12.5 Contact time: 15 min.
4.	Rinse	60°C water; hosing until all detergent is removed

(cont.)

Table 5.19. (Cont.)

CLEANING AND DISINFECTION

INSTRUCTION NO.: P 18 10 05 ***REVISION NO.:*** **3** ***DATE:*** **1993-03-16**

1. EDITION: 1993-01-15 ***PAGE:*** 1 ***of*** 2

5.	Visual inspection	All cleaned surfaces are inspected. If residues are observed steps 2, 3 and 4 are repeated. The inspection is recorded in the Logbook of Hall C.
6.	Disinfection	Chlorine (YY) is applied to all surfaces. Dosage: 1 l to 50 l of cold water. Free chlorine level: > 200 ppm. Contact time: 10-15 min.
7.	Rinse	Hosing with cold water
8.	Inspection	Before start of production a visual inspection is carried out and the result is recorded in the logbook of the Process Hall C.
9.	Reporting	Inspection result are reported by the Foreman to the Technical Director, who decides on the corrective actions to be initiated.

ISSUED BY: ***APPROVED BY:***

Table 5.20. Overview of the categories of documents included in a Quality System.

DOCUMENTS

SYSTEM DOCUMENTATION	DOCUMENTATION OF RESULTS	SPECIFICATIONS
- QUALITY MANUAL - PROCEDURES - INSTRUCTIONS - FORMS	- INSPECTIONS - ANALYTICAL CONTROL - PROCESS CONTROL - NON-CONFORMANCE - CORRECTIVE ACTION	- RECIPES - QUALITY - SPECIFICATIONS - PROCESS- SPECIFICATIONS

5.2.4. Establishment and implementation of Quality System

The work involved in establishing and implementing a Quality System e.g. ISO 9001 or 9002 should not be underestimated. It is a very demanding project both in terms of man-hours and of resources. Proper planning including well defined project organisation and very often with assistance from outside consultants is a necessity for a successful result. Further, full commitment and motivation as well as intensive training of all employees are indispensable requirements.

Table 5.20 and Figure 5.5 illustrate the various activities involved as well as a time schedule for a smaller company. To initiate the project and as responsible for its completion, a Quality Management Group will normally be formed. For food industries this Group may comprise the following persons: Managing Director, Technical Director, R & D Manager, Sales Manager and Head of Laboratory. The key functions of the Group can be summarized as follows:

- Definition of quality policy and objectives.
- Definition of responsibilities.
- Decision on time schedule for the project from start to certification.
- Allocation of resources required.
- Information and motivation of all staff members.
- Training of all employees.
- Follow-up on time schedules.
- Resolving differences of opinions, argumenting etc.

The various phases and activities following the formation of the Quality Management Group are shown in Table 5.21 and Figure 5.5 which almost speak for themselves. The time required i.e. about 1-2 years or more, for the implementation and certification of the system in a medium size company, should be noticed.

Table 5.21. Phases of a quality system approach.

Formation of a quality management group
Hazard Analysis
Audit of present system elements
Estimation of resources and total period of time required for the project including certification
Formation of project organisation
Preparation of Quality Manual (level 1)
Training of all staff members
Definition of procedures and instructions (levels 2 and 3) to be included in departmental manuals i.e. table of contents
Decision on time schedule for preparation of departmental manuals
Establishment of working groups for the preparation of the individual procedures and instructions
Commenting, reviewing, approving and issuing procedures and instructions
Implementation of procedures and instructions
First approach to certifying body
Internal auditing
Corrections, adjustments etc.
Further training of staff
Certification

Hazard analysis.	4 weeks
Audit of existing system/elements.	20 weeks
Preparation of Quality Manual.	6 weeks
Information and training of employees.	5 weeks; Contact to certifying body
Formation of working groups, preparation of procedures and instructions.	30 weeks
Adjustments. Implementations. Further training of employees. Internal audits.	10 weeks
Final Adjustments.	2 weeks
Certification.	4 weeks
	0 5 25 30 35 40 45 50 55 60 65 70 75 80 Weeks

Figure 5.5. Time schedule for establishing and implementing a quality system in a small size food processing plant.

5.2.5. Advantages and disadvantages experienced by ISO 9000 certified companies

An analysis encompassing one hundred ISO 9000 certified companies showed that they all had experienced significant advantages. Marketing merits, reduced quality costs, and higher efficiency were the main advantages mentioned, all contributing to a higher profitability. Findings which agree very well with the general opinion within the food industry in Europe. As regards quality costs, Figure 5.6 shows how profit is typically created when Quality Management is implemented in a company. The reduction in quality costs observed in practice can be as much as 5-15% of the company turnover, and investment in Quality Management have proved to be profitable.

The disadvantages experienced appear to be too much bureaucracy and lack of flexibility, which are inherent parts of the ISO standards, together with the significant amount of paper work involved.

The main objective of Quality Management according to the ISO 9000 series can be defined as meeting the agreed requirements of the customer. This underlines that the quality of a company's products is the key factor in the performance of the company. ISO 9000 is clearly a system seeing quality from the industry's point of view.

The response from the food industry has been slow compared to other industries. However, a rapidly increasing interest is now seen in Denmark and several other European countries. The interest is not limited to food processing plants; all links from primary production to the final product are becoming involved. One may expect that in the near future the whole chain from the primary producer to consumers will be covered by certified quality systems. Projects on certification of farms are in progress in Denmark, and fishing boats have already been certified according to ISO 9000.

This development will form a proper background for meeting the worldwide trend towards more stringent customer expectations.

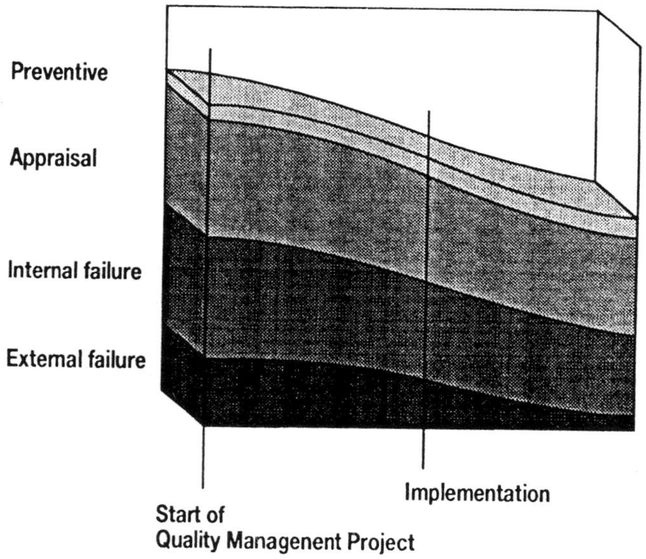

Figure 5.6. Economical benefits from introducing a quality system.

6. CLEANING AND SANITATION IN SEAFOOD PROCESSING

6.1. WATER QUALITY IN PROCESSING AND CLEANING PROCEDURES

This Section is prepared by Dr. Susanne Knøchel

6.1.1. Definitions of drinking water quality

Water used for food processing is one of the important critical control points. This is true for water used as an ingredient, for water used as final rinse when cleaning equipment or water which is in any way likely to come into contact with the product. Most often it is just stated that the water should meet drinking water standards and both supply and quality are mostly taken for granted. However, local standards may vary somewhat or may even be absent. The quality of the source water differs enormously from place to place as does the water treatment. The control exerted by the local regulatory authorities may also differ greatly depending on the local situation. Lastly, in-plant problems may sometimes render potable water unfit as drinking water at the final point of use.

So how can acceptable drinking water quality be defined ? What is the rationale behind these guidelines ? And what can the food processors do ?

A universally accepted list of standards for biological and physico-chemical parameters for drinking water does not exist.

WHO issued an excellent book called "Guidelines for drinking water quality", Vol 1, 2, and 3 (WHO 1984b). Volume 1 deals with the guideline values, Volume 2 contains monographs on each contaminant, and Volume 3 gives information on how to handle the water supplies in small, rural communities. In this book, WHO recognizes that very stringent **standards** cannot be used universally as this may severely limit the availability of water and instead, a range of guideline values for more than 60 parameters have been elaborated. A general review of the standards employed by WHO, EEC, Canada and USA is given by Premazzi et al. (1989). It is recognized that e.g. most of the rural wells all over the world would have difficulties meeting all the guideline values suggested. It goes without saying that all the parameters cannot be monitored so selection and priorities must be made based on hazard analysis and feasibility. Most nations (or in some cases even individual provinces) have their own guidelines or standards. The basic microbiological guideline values, however, do not differ so much from place to place. Below are the microbiological parameters and guideline values suggested by WHO (Table 6.1) and EEC (Table 6.2)

Table 6.1. Microbiological criteria (guidelines) for drinking water quality (WHO 1984b).

Organism in 100 ml[1)]	Guideline value	Remarks
Piped water supplies		
Treated water entering the distribution system		
fecal coliforms	0	turbidity < 1 NTU; for disinfection with chlorine, Ph preferably < 8.0, free chlorine residual 0.2-0.5 mg/l following 30 min (minimum) contact
coliform organisms	0	
Water in the distribution system		
fecal coliforms	0	
coliform organisms	0	in 95% of samples examined throughout the year - in the case of large supplies when sufficient samples are examined.
coliform organisms	3	in an occasional sample but not in consecutive samples

1) Multiple tube techniques (MPN procedure) and the membrane filtration technique have been considered as capable of yielding comparable information.

Table 6.2. Microbiological criteria (guidelines) for drinking water quality (EEC 1980).

			Maximum admissible concentration (MAC)	
Parameters	Results: volume of the sample (ml)	Guide level (GL)	Membrane filter method	Multiple tube method (MPN)
Total coliforms	100	-	0	MPN<1
Fecal coliforms	100	-	0	MPN<1
Fecal streptococci	100	-	0	MPN<1
Sulphite-reducing clostridia	20	-	0	MPN<1
Total bacteria counts for water supplied for human consumption	1[1)] 1[2)]	10[1)] 100[2)]		

1) Incubation at 37°C 2) Incubation at 22°C

In the case of water used for food production, it is of vital importance that these microbiological guideline values should be met since potentially pathogenic bacteria are capable of multiplying rapidly if they are introduced into foodstuffs making even initially low and non-infectious doses of bacterial pathogens, a hazard.

Disinfectant residuals should be monitored where possible and periodic verifications of the bacteriological quality should be conducted. Turbidity, colour, taste and odour are also easily monitored parameters. If there are local problems with chemical constituents (e.g. fluoride, iron) or contaminants from industry or agriculture (e.g. nitrate, pesticides, mining waste) these should hopefully be monitored and dealt with by the water suppliers.

6.1.2. Effect of water treatment including disinfection on microbiological agents

Water treatments vary from region to region depending on the water sources available. While groundwater from sedimentary aquifers has undergone extensive filtration the water from hard rock aquifers or surface water sources should be filtered as part of the water treatment in order to decrease the content of particulates, microorganisms and organic and inorganic matter.

Parasites are removed to a large extent by filtration. The levels of bacteria and virus also decrease markedly and the removal mechanisms are both filtration and adsorption. The cation concentration influences adsorption, i.e., increasing concentrations give rise to increased adsorption. Ca^{2+} and Mg^{2+} seem to be especially efficient. These small cations will decrease the repulsive forces between the soil particles and the microorganisms. Iron oxides also have a high affinity for viruses as well as bacteria. Ferric hydroxide impregnated lignite has even been suggested as a local filtration/adsorption media (Prasad and Chaudhuri 1989).

The disinfection efficiency is greatly affected by **type of disinfectant**, **type and state of microorganism**, **water quality parameters such as turbidity** (or suspended solids), **organic matter**, **some inorganic compounds**, **Ph** and **temperature**. The "**hardness**" of the water may indirectly influence disinfection since deposits may harbour microorganisms and protect them from cleaning agents and disinfectants.

Type of disinfectant

By far the most widespread disinfectant is chlorine but also chloramines, chlorine dioxide, ozone and UV light are being used in some instances. **Chlorine** is cheap and available in most places and monitoring free residual levels is simple. It is desirable to maintain a free residual chlorine level of 0.2-0.5 mg/l in the distribution system (WHO 1984b). For sanitation of clean equipment, up to 200 mg/l is used. To avoid corrosion lower concentrations of 50-100 mg/l and longer contact times (10-20 minutes) are often used. **Chloramines** are more stable but less bacteriocidal and much less efficient towards parasites and virus than chlorine. **Chlorine dioxide** is, if anything, more microbicidal than chlorine, especially at high Ph, but there is concern with regards to the by-products. In the case of ozone and UV light there is no residual to monitor. Ozone seems to be very efficient towards protozoa. The efficiency of UV disinfection decreases markedly if there is any turbidity or dispersed organic matter and problems are often encountered due to lack of lamp maintenance.

Type and state of microorganism

In the case of most disinfectants, the order of sensitivity is:

vegetative bacteria > viruses > bacterial spores, acid-fast bacteria and protozoan cysts

The sensitivity varies within groups and even within species. Our indicator bacteria are unfortunately among the more sensitive microorganisms and the presence of e.g. fecal coliforms in treated, disinfected water is therefore a very clear indication that the water contains potentially pathogenic microorganisms while the absence of such indicator bacteria do not guarantee pathogen-free water.

Bacteria from nutrient-poor media as well as otherwise **stressed bacteria** may also exhibit greatly increased resistance. Some of the effects mentioned on the efficiency of free chlorine are illustrated in Table 6.3.

Water quality factors

If microbes are associated with **granular material or other surfaces** the effect of a disinfectant such as chlorine decreases drastically. Attachment of *Klebsiella pneumonia* to glass surfaces may for example increase the resistance to free chlorine 150-fold (Sobsey 1989).

Organic matter may react and "consume" disinfectants such as chlorine and ozone and the presence will also interfere with UV light. The chloramines are less susceptible to organic matter.

Ph is important in disinfection with chlorine and chlorine dioxide with greater inactivation at low Ph in the case of chlorine and greater inactivation at high Ph in the case of chlorine dioxide (Sobsey 1989).

In general, higher **temperatures** result in increased inactivation rates.

Table 6.3. Inactivation of microorganisms by free chlorine.

Organism	Water	Cl_2 residues, mg/l	Temperature, °C	Ph	Time, min.	Reduction %	C*t[1)]
E. coli	BDF[2)]	0.2	25	7.0	15	99.997	ND[3)]
E. coli	CDF[4)]	1.5	4	?	60	99.9	2.5
E. coli + GAC[5)]	CDF	1.5	4	?	60	<< 10	>> 60
L. pneumophila (water grown)	tap	0.25	20	7.7	58	99	15
L. pneumophila (media grown)	tap	0.25	20	7.7	4	99	1.1
Acid-fast *Mycobacterium chelonei*	BDF	0.3	25	7.0	60	40	>> 60
Virus							
Hepatitis A	BDF	0.5	5	10.0	49.6	99.99	12.3
Hepatitis A	BDF	0.5	5	6.0	6.5	99.99	1.8
Parasites							
G. lamblia	BDF	0.2-0.3	5	6.0	-	99	54-87
G. lamblia	BDF	0.2-0.3	5	7.0	-	99	83-133
G. lamblia	BDF	0.2-0.3	5	8.0	-	99	119-192

1) C*t product of disinfectant concentration (C) in mg/l and contact time (t) in minutes for 99% inactivation (mod.a. Sobsey 1989)
2) BDF = buffered demand free
3) ND = no data
4) CDF = Chlorine demand free
5) GAC = granular activated carbon

6.1.3. Use of non-potable water in a plant

The use of non-potable water may be necessary for water conservation purposes or desirable because of cost. The water may e.g. be surface water, sea water or chlorinated water from can cooling. Relatively clean water such as chlorinated water from can cooling operations may be used for washing cans after closing before heat treatment, for transporting raw materials before processing (after the water has cooled off), for initial washing of boxes, for cooling of compressors, for use in fire protection lines in non-food areas and for fuming of waste material. It is absolutely necessary that potable and non-potable water should be in separate distribution systems which should be clearly identifiable. If potable water is used to supplement a non-potable supply the potable source must be protected against valve leaking, back-pressure e.g. by adequate air-gaps (Katsuyama and Strachan 1980). Back-flow due to sudden pressure differentials or blockage of pipes have unfortunately occurred in many systems.

Potentially contaminated water such as coastal water or surface water should not be used at the production premises but may, if aesthetically acceptable, be used for removing waste material in places where no contact to food is possible.

6.1.4. A water quality monitoring system

The responsible person should have continuously updated reference drawings of the pipe system and the authority to remove dead-ends. Especially in cases where a plant has undergone many changes, the piperuns may become more and more complicated over the years. The person should also be in contact with the local waterworks and the authorities in order to be informed of special events (repairs, pollution accidents or other changes).

A quality monitoring scheme could consist of a schematizised plan of all the sampling points and a checklist for each point describing what to examine and why, the frequency, who takes the sample, who does the analysis, what is the limit (value, tolerance) and what to do in case of deviation (Poretti 1990). If the water is obviously polluted there is of course no reason to wait for analytical results. The sampling frequency and the range of parameters will vary with the circumstances and the needs and possibilities of the specific plant. A minimum program may for example consists of monitoring free chlorine daily and total counts plus coliforms on a weekly basis and a special, more intense monitoring program to be used after repairs, when using new water supplies etc.

The technical procedures describing the analyses for the common indicator organisms are given in standard textbooks. The WHO "Guidelines for drinking-water Quality", vol. 3 (WHO 1984b) mentions some methods and equipment suitable for small, rural supplies. The values used by the company should refer to the specific method employed and the recommendations should include how to sample (tap flow, volume, sampling vessel, labelling etc.) and how to handle and examine the sample. Even though the commonly used methods for detecting e.g. fecal coliforms are standard analyses faulty handling of the samples often occurs. Samples should be processed within 24 hours or less and be kept cool, but not frozen (preferably below 5°C), and in the dark. The impact of sunlight can be very dramatic causing false negative results (Knøchel 1990).

If chlorination is used for disinfection, monitoring of the free chlorine level is the simplest way of checking the water treatment and should be performed most often (e.g. on a daily basis). Simple laboratory methods are described by WHO (1984b) and commercial dip-sticks are now available for on-the-spot measurements (e.g. Merckoquant Chlor 100 from Merck). The microbiological indicator parameters may be checked less frequently. If disinfection systems leaving no residuals are being used, checking the equipment should be done regularly. The performance of the systems may be monitored at weekly intervals using indicator bacteria measurements.

6.2. CLEANING AND DISINFECTION

This Section is prepared by Professor Mogens Jakobsen

6.2.1. Introduction

Cleaning and disinfection belong to the most important operations in today's food industries. Numerous and costly cases of food spoilage and unacceptable contamination with pathogenic bacteria has been traced back to failures or insufficiencies of these procedures.

The standards of hygiene required to avoid such problems are variable. In a plant, packaging products processed for safety (e.g. by heat treatment) requirements will be very strict whereas handling of fresh chilled fish with a short shelf life and which is cooked before consumption, will be less demanding.

Factors like housekeeping, personal hygiene, training and education, plant layout, design of equipment and machines, characteristics of materials selected, the maintenance and general condition of the plant can easily become more important than the actual cleaning and disinfection. For optimal use of resources and to ensure the microbiological quality of foods, it is important that all such factors are addressed when deciding on cleaning and disinfection procedures.

In some cases it may even be best to avoid cleaning and disinfection, because more harm than good can be done. As an example, this applies for dust accumulated on pipes and constructions unless time allows for a complete removal. Further, as another example, dry areas should always be kept dry and cleaning will then be limited to vacuuming if available, or sweeping, brushing etc.

It follows from the above that for each particular food plant or operation, implementation of cleaning and disinfection procedures is a project on its own where specialists, internal or external, should be consulted.

Cleaning and disinfection will be processes like any other plant operation, and they should be equally documented and so should the corresponding process control i.e. the control of cleaning and disinfection respectively. If a HACCP concept is applied, these procedures should be treated as Critical Control Points (CCPs). If a Quality System like ISO 9000 is in operation, they should be integrated in the System as illustrated in the previous chapter of this book. Responsible management realizes that these procedures are integrated parts of production and poor hygienic condition in food processing plants will primarily be caused by management lack of knowledge and commitment.

For the whole process, three distinct operations are involved i.e.

i) preparatory work; ii) cleaning and iii) disinfection. They are clearly distinct operations but linked firmly together in the way that the final result will not be acceptable, unless all three are carried out correctly. Table 6.4 shows the various steps, which will be included in a complete cycle.

Table 6.4. Steps included in the complete cycle of preparatory work, cleaning, disinfection and control.

1	Remove food products, clear the area for bins, containers etc.
2	Dismantle equipment to expose surfaces to be cleaned. Remove small equipment, parts and fittings to be cleaned in a specified area. Cover sensitive installations, to protect them against water etc.
3	Clear the area, machines and equipment for food residues by flushing with water (cold or hot) and by using brushes, brooms etc.
4	Apply the cleaning agent and use mechanical energy (e.g. pressure and brushes) as required.
5	Rinse thoroughly with water to completely remove the cleaning agent after the appropriate contact time, (residues may completely inhibit the effect of disinfection).
6	Control of cleaning.
7	Sterilization by chemical disinfectants or heat.
8	Rinse the sterilant off with water after the appropriate contact time. This final rinse is not needed for some sterilants e.g. H_2O_2 based formulations which decompose rapidly.
9	After the final rinse, equipment is reassembled and allowed to dry.
10	Control of cleaning and disinfection.
11	In some cases it will be good practice to re-disinfect (e.g. with hot water or low levels of chlorine) just before production starts.

6.2.2. Preparatory work

In this phase, the processing area is cleared of remaining products, spills, containers and other loose items. Machines, conveyors etc. are dismantled so that all locations, where microorganisms can accumulate become accessible for cleaning and disinfection. Further electrical installations and other sensitive systems should be protected against water and the chemicals used.

Before use of the cleaning agent, a gross food debris removal procedure should be carried out by brushing, scraping or similar. All surfaces should be further prepared for the use of cleaning agents by a pre-rinse activity preferably with cold water which will not to coagulate proteins. Hot water may be used to remove fat or sugars in cases, where protein is not present in significant amounts.

Completion of the preparatory work should be checked and recorded as any other process, to ensure the quality of the complete cycle of cleaning and disinfection.

6.2.3. Cleaning

Cleaning is undertaken to remove all undesirable materials (food residues, microorganisms, scales, grease etc.) from the surfaces of the plant and the process equipment, leaving surfaces clean, as determined by sight and touch and with no residues from cleaning agents.

Microorganisms present will either be incorporated in the various materials or they attach to the surfaces as biofilms. The latter will not be removed completely by cleaning, but experience has shown that a majority of the microorganisms will be removed. However, there will still be some left to be inactivated during the disinfection.

The effectiveness of a cleaning procedure in general depends upon:

- The type and amount of material to be removed.

- The chemical and physio-chemical properties of the cleaning agent (such as acid or alkali strength, surface activity, etc.) at the concentration, temperature and exposure time used.

- The mechanical energy applied e.g. turbulence of cleaning solutions in pipes, stirring effect, impact of water jet, "elbow-grease", etc.

- Condition of the surface to be cleaned.

Some surfaces e.g. corroded steel and aluminum surfaces can simply not be cleaned which means that disinfection also becomes very inefficient. The same applies for other surfaces e.g. wood, rubber etc. The preferred material obviously will be high quality stainless steel.

The types of residues to be removed in food plants, will mainly be the following:

- Organic matter, such as protein, fat and carbohydrate. These are most effectively removed by strongly alkaline detergents (especially caustic soda, NaOH).
  Further, it is found that combinations of acid detergents (especially phosphoric acid) and non-ionic surfactants are effective against organic matter.

- Inorganic matter, such as salts of calcium and other metals. In beer stone, milk stone, etc., salts are encrusted with protein residues.
  These are most effectively removed by acid cleaning agents.

- Biofilms, formed by bacteria, moulds, yeast and algae can be removed by cleaning agents that are effective against organic matter.

Most cleaning agents work faster and more effectively at higher temperatures, so it can be profitable to clean at a high temperature. Cleaning is often carried out at 60 - 80°C in areas where it pays energy wise to use such high temperatures.

Water

Water is used as a solvent for all cleaning and sterilizing agents, and also for intermediate rinses and final rinse of equipment.

The chemical and microbiological quality of the water is therefore of decisive importance for the efficiency of the cleaning procedures as already described in a previous section of this chapter. In principle, water used for cleaning must be potable.

Hard water contains a large amount of calcium and magnesium ions. When the water is heated, calcium and magnesium salts corresponding to the temporary hardness will precipitate as insoluble salts. Also, some cleaning agents, especially alkalis, can precipitate calcium and magnesium salts.

Apart from reducing the effectiveness of detergents hard water leads to the formation of deposits or scales. Scales which can be formed in several other ways are not only unsightly but objectionable of several reasons:

- They harbour and protect microorganisms.

- They reduce the rate of heat exchange on heat exchanger surfaces. This could lead to underprocessing, underpasteurization or understerilization.

- The presence of scales tends to increase corrosion.

The formation of scales can be reduced by addition of chelating and sequestering agents, which bind calcium and magnesium in insoluble complexes. However, it is advisable to prevent precipitations by softening the water before it is used for cleaning. Softening can be effectively achieved by ion exchange, in which the calcium and magnesium ions are replaced by sodium ions, the salts of which are soluble. A modern, and more costly, method of softening water is by means of reverse osmosis.

Microbiological purity of water to be used for final rinse must be beyond reproach. If not, it will in some cases be acceptable to include low levels of chlorine i.e. a few ppm.

Cleaning agents

The ideal detergent would be characterized by the following properties:

- It possesses sufficient chemical power to dissolve the material to be removed.

- It has a surface tension low enough to penetrate into cracks and crevices; it should be able to disperse the loosened debris and hold it in suspension.

- If used with hard water, it should possess water softening and calcium salt dissolving properties to prevent precipitation and build-up of scale on surfaces.

- It rinses freely from the plant, leaving this clean and free from residues, which could harm the products and affect sterilization negatively.

- It does not cause corrosion or other deterioration of the plant. It is recommended always to check by consulting the supplier of machines etc.

- It is not hazardous for the operator.

- It is compatible with the cleaning procedure being used, whether manual or mechanical.

- If solid, it should be easily soluble in water and its concentration easily checked.

- It complies with legal requirements concerning safety and health as well as biodegradability.

- It is reasonably economical to use.

A detergent with all these characteristics does not exist. So one must, for each individual cleaning operation, select a compromise by choosing a useable cleaning agent and water treatment additives so that the combined detergent has the properties that are most important for the procedure concerned.

When choosing a cleaning agent, one can pick either a ready-mixed factory product, which has the desired properties, or it can be homemade following the guidelines given in Table 6.5. In this case it must be assured that the components are mutually compatible.

Table 6.5 (from Lewis 1980) shows important characteristics of the cleaning agents most commonly used in the food industry.

Table 6.5. Types, functions, and limitations of cleaning agents used in the food industries (from Lewis 1980).

Categories of aqueous cleaners	Approximate concentrations for use (%, w/v)[1]	Examples of chemical used[2]	Functions	Limitations
Clean water	100	Usually contains dissolved air and soluble minerals in small amounts	Solvent and carrier for soils, as well as chemical cleaners	Hard water leaves deposit on surfaces. Residual moisture may allow microbial growth on washed surfaces.
Strong alkali	1 - 5	Sodium hydroxide Sodium orthosilicate Sodium sesquisilicate	Detergents for fat and protein. Precipitate water hardness	Highly corrosive. Difficult to remove by rinsing. Irritating to skin and mucous membranes.
Mild alkali	1 - 10	Sodium carbonate Sodium sesquisilicate Trisodium phosphate Sodium tetraborate	Detergents. Buffers at Ph 8.4 or above Water softeners	Mildly corrosive. High concentrations are irritating to skin
Inorganic acid	0.5	Hydrochloric Sulphuric Nitric Phosphoric Sulphamic	Produce Ph 2.5 or below Remove inorganic precipitates from surfaces	Very corrosive to metals, but can be partially inhibited by anti-corrosive agents. Irritating to skin and mucous membranes
Organic acids	0.1 - 2	Acetic Hydroxyacetic Lactic Gluconic Citric Tartaric Levulinic Saccharic		Moderately corrosive, but can be inhibited by various anti-corrosive compounds

(cont.)

Table 6.5. (cont.) Types, functions, and limitations of cleaning agents used in the food industries (from Lewis 1980).

Categories of aqueous cleaners	Approximate concentrations for use (%, w/v)[1)]	Examples of chemical used[2)]	Functions	Limitations
Anionic wetting agents	0.15 or less	Soaps Sulphated alcohols Sulphated hydrocarbons Aryl-alkyl polyether sulphates Sulphonated amides Alkyl-arylsuphonated	Wet surfaces Penetrate crevices and woven fabrics Effective detergents Emulsifiers for oils, fats, waxes, and pigments Compatible with acid or alkaline cleaners and may be synergistic	Some foam excessively Not compatible with cationic wetting agents
Non-ionic wetting agents	0.15 or less	Polyethenoxyethers Ethylene oxide-fatty acid condensates Amine-fatty acid condensate	Excellent detergents for oil. Used in mixtures of wetting agents to control foam	May be sensitive to acids
Cationic wetting agents	0.15 or less	Quaternary ammonium	Some wetting effect Antibacterial action	Not compatible with anionic wetting agents
Sequestering agents	Variable (depending on hardness of water)	Tetrasodium pyrophosphate Sodium tripolyphosphate Sodium hexametaphosphate Sodium tetrapolyphosphate Sodium acid pyrophosphate Ethylenediaminetetra-acetic acid (sodium salt) Sodium gluconate with or without 3% sodium hydroxide	Form soluble complexes with metal ions such as calcium, magnesium and iron to prevent film formation on equipment and utensils See also strong and mild alkalis above	Phosphates are inactivated by protracted exposure to heat Phosphates are unstable in acid solution

(cont.)

Table 6.5. (cont.) Types, functions, and limitations of cleaning agents used in the food industries (from Lewis 1980).

Categories of aqueous cleaners	Approximate concentrations for use (%, w/v)[1]	Examples of chemical used[2]	Functions	Limitations
Abrasives	Variable	Volcanic ash Seismotite Pumice Feldspar Silica flour Steel wool[3] Metal of plastic 'chlore balls'[3] Scrub brushes	Removal of dirt from surfaces with scrubbing Can be used with detergents for difficult cleaning jobs	Scratch surfaces Particles may become imbedded in equipment and later appear in food Damage skin of workers
Chlorinated compounds	1	Dichlorocyanuric acid Trichlorocyanuric acid Dichlorohydantoin	Used with alkaline cleaners to petizing of proteins and minimize milk deposits	Not germicidal because of high Ph Concentrations vary depending on the alkaline cleaner and conditions of use
Amphoterics	1.2	Mixtures of a cationic amine salt or a quaternary ammonium compound with an anionic carboxy compound, a sulfate ester, or a sulfonic acid	Loosen and soften charred food residues on ovens or other metal and ceramic surfaces	Not suitable for use on food contact surfaces[4]
Enzymes	0.3 - 1	Proteolytic enzymes	Digest proteins and other complex organic soils	Inactivated by heat Some people become hyper-sensitive to the commercial preparations.

1) Concentration of cleaning agent in solution as applied to equipment
2) Some regulatory agencies require prior approval
3) Steel wool and metal 'chlore balls' should **not** be used on food plant
4) Some amphoteric disinfectants are used on food contact surfaces

Cleaning systems

The various steps shown in Table 6.4 and including sterilization, represents the most comprehensive procedure for manual cleaning and disinfection or Clean Out of Place (COP). It is suitable for modern plants. For cleaning liquid handling plants like breweries and dairies Clean In Place (CIP) Systems will be used, based on circulation by pumping of water, cleaning agents and disinfectants. In principle the two systems will be similar.

In most factories, a combination of COP and CIP will be used. Use of CIP may be limited to part of the plants or even to a particular machine. However, regardless of the type and size of food production the general principles behind the complex cycle shown in Table 6.4 should be kept in mind and applied to ensure effective cleaning and disinfection.

The frequency of cleaning and disinfection will vary from several times during the working day i.e. at every major break to once every day, at end of production, or even less frequent. Sometimes disinfection will not be included e.g. in areas to be kept dry and for environments with materials which cannot be disinfected or premises unsuitable for disinfection. In such cases cleaning is still very important for the general appearance and hygienic condition of the plant or premises and the general attitude towards hygiene of the employees.

Control of cleaning

As mentioned earlier, effective cleaning is a prerequisite for an efficient disinfection. This indicates the importance of controlling cleaning. As described in Table 5.18 in the previous Chapter the most important control is visual inspection and other rapid tests to demonstrate the following important results of cleaning:

- That all cleaned surfaces are visibly clean.

- That all surfaces by feeling are free from food residues, scales and other materials and by smelling free from undesirable odours.

Further, the concentrations and Ph-values of cleaning agents, temperatures, if hot cleaning is used, and contact times should be monitored and registered. Ph measurements, or similar testing, of rinse water may be used to ensure that the cleaning agent is removed so that it will not interfere with the disinfectant.

These controls are all rapid and allow immediate decisions to be made as to whether cleaning should be repeated, partly or completely, or to proceed to the process of disinfection. All controls etc. shall be registered as part of the Quality System.

At this stage, microbiological control serves no real purpose. Firstly biofilms and surviving microorganisms are likely to be present and secondly, reliable rapid methods are not available.

6.2.4. Disinfection

Traditionally, the terms "disinfection" and "disinfectants" are used to describe procedures and agents used in food industries to ensure a microbiologically acceptable standard of hygiene. This practice will be followed although it is realized that the procedures and agents described will rarely introduce 'sterility' i.e. total absence of viable micro-organisms.

Disinfection can be effected by physical treatments such as heat, U.V. irradiation, or by means of chemical compounds. Among the physical treatments, only heat shall be described.

The use of heat in the form of steam or hot water is a very safe method and a widely used method of disinfection. The most commonly used chemicals for disinfection are:

- Chlorine and chlorine compounds.
- Iodophors.
- Peracetic acid and hydrogen peroxide.
- Quaternary ammonium compounds.
- Ampholytic compounds.

Table 6.6 summarizes the characteristics for some of these disinfectants and for the use of steam.

Disinfection by use of heat

Heating at suitably high temperatures for a suitably long time is the safest method for killing microorganisms. The velocity with which heat killing occurs depends on temperature, humidity, type of microorganism and the environment in which the microorganisms occur during heat treatment. If microorganisms are entrapped in scales or other substances, they are protected and not even heating may be effective. It is important to recall the kinetics for heat inactivation of microorganisms:

$$\log C_t = \log C_o - K \times t,$$

where C_o = original population of living microorganisms (initial viable count) and C_t = total surviving after time t. K is a constant (= slope of the straight line) and depends on the microorganism concerned and the experimental conditions. K is described as the death rate. It is seen that the number of surviving microorganisms at time 't' is determined by the initial level of infection, as well as the death rate constant and the heating time.

Circulation of hot water (about 90°C) is very effective. The water should be circulated for at least 20 minutes after the temperature of the return water has risen to 85°C or more. Obviously steaming is equally effective when applicable.

Table 6.6. Comparison of the more commonly used disinfectants (ICMSF 1988).

		Steam	Chlorine	Iodophores	QAC/QUATS surfactants	Acid anionic
Effective against	Gram-positive bacteria (lactics, clostridia, *Bacillus*, *Staphylococcus*)	Best	Good	Good	Good	Good
	Gram-negative bacteria (*E. coli*, *Salmonella*, psychrotrophs)	Best	Good	Good	Poor	Good
	Spores	Good	Good	Poor		Fair
	Bacteriophages	Best	Good	Good		Poor
Properties	Corrosive	No	Yes	Slightly	No	Slightly
	Affected by hard water	No	(No)	Slightly	Some are	Slightly
	Irritative to skin	Yes	Yes	Yes	No	Yes
	Affected by organic matter	No	Most	Somewhat	Least	Somewhat
	Incompatible with:	Materials sensitive to high temperature	Phenols, amines, soft metals	Starch, silver	Anionic wetting agents, soaps.	Cationic surfactants and alkaline detergents
	Stability of use solution		Dissipates rapidly	Dissipates slowly	Stable	Stable
	Stability in hot solution (greater than 66°C)		Unstable, some compounds stable	Highly usable (best used below 45°C)	Stable	Stable
	Leaves active residue	No	No	Yes	Yes	Yes
	Tests for active residue chemical	Unnecessary	Simple	Simple	Simple	Difficult
	Maximum level permitted by USDA and FDA w/o rinse	No limit	200 ppm	25 ppm	25 ppm	
	Effective at neutral Ph	Yes	Yes	No	No	No

Disinfection by use of chemical agents

With the use of chemical disinfectants, the death rate for microorganisms depends, among other things, upon the agent's microbicidal properties, concentration, temperature and Ph as well as the degree of contact between disinfectant and microorganisms. Good contact is obtained, e.g. by stirring, turbulence, smooth surfaces and low surface tension. As with heat disinfection, different microorganisms show different resistance to chemical sterilants. It is also so that contamination by inorganic or organic matter can reduce the death rate considerably. As mentioned before an effective disinfection can only be obtained after an effective cleaning. The desirable plant disinfectant would be characterized by the following properties:

- It has sufficient anti-microbial effect to kill the microorganisms present in the available time and should have a sufficiently low surface tension to ensure good penetration into pores and cracks.

- It rinses freely from the plant, leaving this clean and free from residues which could harm the products.

- It must not lead to development of resistant strains or any surviving microorganisms.

- It does not cause corrosion or other deterioration of the plant. It is recommended that the suppliers of machines etc. be asked before chlorine or other aggressive disinfectants are taken into use.

- It is not hazardous to the operator.

- It is compatible with the disinfection procedure being used, whether manual or mechanical.

- If solid, it should be easily soluble in water.

- Its concentration is easily checked.

- It is stable for extended storage periods.

- It complies with legal requirements concerning safety and health as well as biodegradability.

- It is reasonably economical in use.

It will often be necessary to combine sterilants with additives in order to obtain the required properties.

To prevent development of resistant strains of microorganisms it can be advantageous to change from one type of sterilant to another from time to time.

This is especially advisable when quaternary ammonium compounds are used.

Among the most used sterilants the following shall be described briefly.

Chlorine is one of the most effective and widely used disinfectants. It is available in several forms like sodium hypochlorite solutions, chloramines and other chlorine containing organic compounds. Gaseous chlorine and chlorine dioxide are also used.

Chlorinated sterilants at a concentration of 200 ppm free chlorine are very active and also with some cleaning effect. The disinfectant effect is considerably decreased when organic residues are present.

The compounds dissolved in water will produce hypochlorous acid, HOCl, which is the active sterilizing agent, acting by oxidation. In solution it is very unstable, particularly in acid solution where oxic chlorine gas will be liberated. Furthermore, solutions are more corrosive at low Ph.

Unfortunately, the germicidal activity is considerably better in acid solution than in alkaline, thus the working Ph should be chosen as a compromise between efficiency and stability. Organic chlorinated sterilants are generally more stable but require longer contact times.

When used in the proper range of values (200 ppm free chlorine), chlorinated sterilants in solutions at ambient temperatures are non-corrosive to high quality stainless steel but they are corrosive to other less resistent materials.

Iodophors contain iodine, bound to a carrier, usually a non-ionic compound, from which the iodine is released for sterilization. Normally the Ph is brought down to 2-4 by means of phosphoric acid. Iodine has its maximum effect at this Ph range.

Iodophors are active disinfectants with broad antimicrobial spectrum just like chlorine. They are inactivated by organic material. Concentrations corresponding to approx. 25 ppm free iodine will be effective.

Commercial formulations are often acidic making them able to dissolve scales. They can be corrosive depending on the formulation and they should not be used above 45°C as free iodine may be liberated. If residues of product and caustic cleaning agents are left in dead legs and similar places, this may in combination with iodophors cause very unpleasant "phenolic" off-flavours.

Hydrogen peroxide and **peracetic** acid are effective sterilants acting by oxidation and with a broad antimicrobial spectrum. Diluted solutions may be used alone or in combination for disinfection of clean surfaces. They lose their activity more readily than other sterilants in the presence of organic substances and they rapidly loose their activity with time.

Quaternary ammonia compounds are cationic surfactants. They are effective fungicides and bactericides but often less effective against Gram negative bacteria. To avoid development of resistant strains of microorganisms, these compounds should only be used alternating with the use of other types of disinfectants.

Due to their low surface tension, they have good penetrating properties and for the same reason, they can be difficult to rinse off.

If quats come into contact with anion-active detergents, they will precipitate and become inactivated. Mixing or successive use of these two types of chemicals must therefore be avoided.

Ampholytic sterilants have properties similar to quaternary ammonia compounds.

Control of disinfection

Control of disinfection will be the final control of the complete cycle of cleaning and disinfection. Provided cleaning has been controlled effectively as described above control of disinfection will be effective when the following conditions are met:

- Control of time and temperature conditions for disinfection by heat.
- Control of active concentrations of chemical disinfectants.
- Control that all surfaces to be disinfected are covered by the disinfectant.
- Control of contact time.

The above controls should be documented and the observations reported and registered as required in standard Quality Systems.

Microbiological testing and control serve the purpose of verification. Various techniques are available but none are ideal and they are not "real time" methods which is highly desirable for control of cleaning and disinfection. Overnight incubation is too late to correct critical situations.

However, if conducted at regular intervals and planned to cover all critical points, useful information from microbiological control can be accumulated with time. Various methods are used and shall be mentioned briefly.

- Swab testing. This is the most usual technique and one of the better ones. By use of a sterile swab of cottonwool, part of the disinfected surface is swabbed, and the bacteria transferred to the swab is transferred to a diluent for determination of colony forming units in standard agar substrates. Swabs are especially useful in places, where other control methods can only be used with difficulty i.e. pockets, valves etc.

- Final rinse water. Membrane filtration of rinse water and incubation on agar substrate is a very sensitive technique for control of CIP systems as well as other cleaning and disinfection systems, where a rinse can be applied.

- Direct surface plates. In these methods petri dishes or contact slides with selective or general purpose agar media are applied to the surface to be examined, followed by incubation and counting of colony forming units. These techniques can only be applied to plane surfaces, which is a limiting factor.

- Bioluminometric assay of ATP. This is almost a "real time" method giving the answer within minutes. It is very sensitive and can be combined with swabbing for collection of microorganisms from surfaces. The method is rather nonspecific, and it may not be able to distinguish between microorganisms and food residues. However, if applied under defined conditions it may prove useful and superior to the conventional methods, because it provides the answer in minutes.

Regardless of the technique used, it is valuable to know from the verification analyses that the system was working, when it was established. There is also a value in knowing trends as expressed in the verification results recorded. The objective of studying trends and conducting the microbiological control of cleaning and disinfection, obviously will be, to take corrective action before loss of control of products or processes occur.

7. ESTABLISHMENTS FOR SEAFOOD PROCESSING

In this chapter some requirements for establishments for (sea)food processing will be discussed. A number of books (Shapton and Shapton 1991, Hayes 1985, ICMSF 1988) and official regulations (e.g. EEC 1991b) give detailed information on requirements for buildings, equipment and processing procedures and these should be consulted if new establishments are being constructed. Some of the more important aspects are considered below.

7.1. PLANT LOCATION, PHYSICAL ENVIRONMENT AND INFRASTRUCTURE

Early considerations in building a new plant is the identification of a suitable location. A number of factors should be considered such as physical, geographical and infrastructure available.

A plant must be located on a plot of adequate size (for present needs and future developments), with easy access by road, rail or water. An adequate supply of potable water and energy must be available throughout the year at a reasonable cost. Special considerations must be given to waste disposal. Seafood processing plants usually contain significant amounts of organic matter which must be removed before waste water is discharged into rivers or the sea. Also solid waste handling needs careful planning, and suitable space, away from the plant, must be allocated or be available.

Assessment of pollution risk from adjacent areas must also be considered. Contaminants such as smoke, dust, ash, foul odours (e.g. neighbouring fish meal plant using poor raw material) are obvious, but even bacteria may have to be considered as airborne contaminants (e.g. proximity of a poultry rearing plant upwind may be a source of *Salmonella* sp).

The immediate physical surroundings of a seafood factory should be landscaped and present on attractive view to the visitor (- or potential buyer of products). However, this should be done in a way so rodents and birds are not attracted. Shrubbery should be at least 10 m away from buildings and a grassfree strip covered with a layer of gravel should follow the outer wall of buildings. This allows for thorough inspection of walls and control of rodents.

7.2. BUILDINGS, CONSTRUCTION AND LAYOUT

A food processing plant shall provide (quoted from Troller 1983):

- Adequate space for equipment, installations and storage of materials.
- Separation of operations that might contaminate food.
- Adequate lightning and ventilation.
- Protection against pests.

The requirements of external walls inclusive roofs, doors and windows are that they should be water-, insect- and rodent proof. Internal walls, on the other hand, should be smooth, flat, resistant to wear and corrosion, impervious, easily cleanable and white or light coloured. Also the floors should ideally be impervious to spillage of product, water and disinfectants, durable to impact, resistant to disinfectants and chemicals used, slip resistant, non-toxic, non-tainting and of good appearance and easy repairable. Floors should be provided with slope to drains to prevent formation of puddles. The technical requirements, choice of materials, cost etc. to obtain these goals may be found in a number of publication such as Shapton and Shapton (1991), Imholte (1984), Troller (1983).

The general layout and arrangements of rooms within a processing establishment is important in order to minimize the risk of contamination of the final product. A large number of bacteria (pathogens and spoilage bacteria) enter with the raw material. To avoid cross contamination, it is therefore essential that raw material is received in a separate area and stored in a separate chillroom. From here the sequence of processing operations should be as direct as possible - and a "straight line" process flow is regarded as the most efficient (Hayes 1985). This layout minimizes the risk of recontamination of a semi-processed product.

Figure 7.1. Clean, orderly and possibly landscaped outside appearance of a food plant is the first impression to a visitor.

A clear physical (e.g. a wall) segregation between "clean" and "unclean" areas is of prime importance. "Unclean" areas are those where raw material is handled and often a cleaning operation (wash) or e.g. a heat treatment (cooking of shrimp) is marking the point, where the process flow goes from "unclean" to "clean" areas. Thus a "clean" area is defined by ICMSF (1988) as an area where any contaminant added to the product will carry over to the final product, i.e. there is no subsequent processing step that will reduce or destroy contaminating microbes. Other terminologies used for "clean" areas are "High Risk Areas" or "High Care Areas".

Also cooled rooms must be separated from hot rooms where cooking, smoking, retorting etc. are taking place. Dry rooms must be separated from wet rooms and ventilation must be sufficient to remove excess humidity.

The separation between the clean and unclean areas must be complete. There should be no human traffic between these areas, and equipment and utensils used in the unclean areas should never be used in the clean area. This means that there should also be separate wash and hygiene facilities for equipment and personnel in these areas. For easy identification the personnel should wear different coloured protective clothing for different operations (e.g. white in the clean area and blue in the unclean).

Equally important in layout and design of food factories is to ensure that there are no interruptions and no "dead ends" in the product flow, where semiprocessed material can accumulate and remain for a long time at ambient temperature. Time/temperature conditions for products during processing are extremely important critical control points (CCPs) in order to prevent bacterial growth. This means that a steady and uninterrupted flow of **all** products is necessary in order to have full control of this critical factor. If any delays in product flow are necessary, the products should be kept chilled.

In addition to facilitate product flow, the factory layout and practices should ensure that:

- All functions should proceed with a no of criss-crossing and backtracking.

- Visitors should move from clean to unclean areas.

- Ingredients should move from "'dirty" to "clean" areas as they become incorporated into food products.

- Conditioned (e.g. chilled) air and drainage should flow from "clean" to "dirty" areas.

- The flow of discarded outer packing material should not cross the flow of either: unwrapped ingredients or finished product.

- There is sufficient space for plant operations including processing, cleaning and maintenance. Space is also required for movement of materials and pedestrians

- Operations are separated as necessary. There are clear advantages in minimizing the number of interior walls since this simplifies the movement of materials and employees, makes supervision easier, and reduces the area of wall that needs cleaning and maintenance (the list is partly after Shapton and Shapton 1991).

Some of the principal requirements for an ideal establishment are outlined in Figure 7.2.

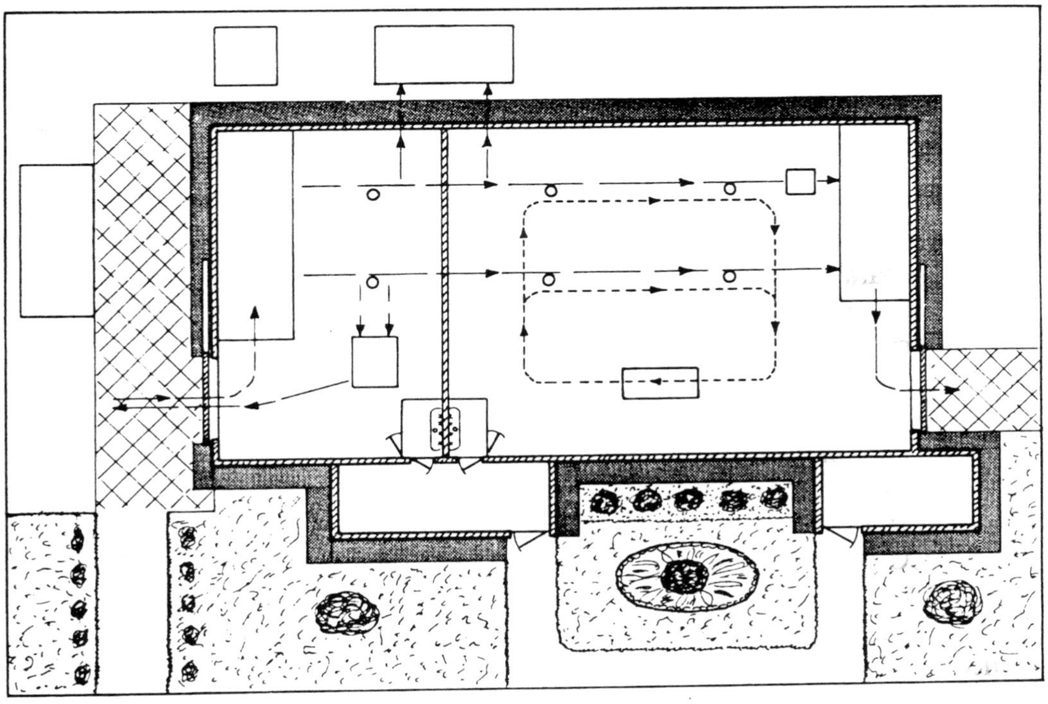

Figure 7.2. Simplified factory layout

7.3. UTENSILS AND EQUIPMENT

A great variety of utensils and equipment is used in the fish industry. There is an abundance of advice and regulations available concerning the requirements for equipment. All of them agree that the food equipment should be non-contaminating and easy to clean. However, the degree of stringency in hygienic requirements must be related to the product being processed. Raw fish for example, do not require the same standard of hygiene as cooked and peeled shrimp. Criteria for hygienic design are particularly important for equipment used in the later stages of processing and particularly after a bacteria-eliminating processing step.

There are seven basic principles for hygienic design agreed upon by a working party appointed by Food Manufacturers Federation (FMF) and Food Machinery Association FMA (FMF/FMA 1967) as quoted by Hayes (1985):

1. All surfaces in contact with food must be inert to the food under the conditions of use and must not migrate to or be absorbed by the food.

2. All surfaces in contact with food must be smooth and non-porous so that tiny particles of food, bacteria, or insect eggs are not caught in microscopic surface crevices and become difficult to dislodge, thus becoming a potential source of contamination.

3. All surfaces in contact with the food must be visible for inspection or the equipment must be readily disassembled for inspection, or it must be demonstrated that routine cleaning procedures eliminate possibility of contamination from bacteria or insects.

4. All surfaces in contact with food must be readily accessible for manual cleaning, or if not readily accessible, then readily disassembled for manual cleaning, or if clean-in-place techniques are used, it must be demonstrated that the results achieved without disassembly are the equivalent of those obtained with disassembly and manual cleaning.

5. All interior surfaces in contact with food must be so arranged that the equipment is self emptying or self draining.

6. Equipment must be so designed as to protect the contents from external contamination.

7. The exterior or non-product contact surfaces should be arranged to prevent harbouring of soils, bacteria or pests in and on the equipment itself as well as in its contact with other equipment, floors, walls or hanging supports.

In the design and construction of equipment it is important to avoid dead areas where food can be trapped and bacterial growth take place. Also dead ends (e.g. thermometer pockets, unused pipe work T-pieces) must be avoided, and any piece of equipment must be designed so the product flow is always following the "first in first out" principle.

Cleanability of equipment involves a number of factors such as construction materials, accessibility and design. The most common design faults which cause poor cleanability are (Shapton and Shapton 1991):

- Poor accessibility (- equipment should be placed at least 1 m from wall, ceiling or nearest equipment).

- Inadequately rounded corners (minimum radius should be 1 cm, but 2 cm is regarded as optimum by the American 3-A Sanitary Standards Committee (Hayes 1985).

- Sharp angles.

- Dead ends (including poorly designed seals).

One general problem of food processing involves the extremes of temperature, abundant use of water, condensation and contamination of food from overhead pipes and surfaces. Equipment design must consider this and include proper protection.

Equipment design is one of the major problems in modern food hygiene. A great number of new machines and equipment are designed and constructed without proper attention to the fact that these tools have to be cleaned and sanitized. The EEC Directive 89/392/EEC (EEC 1989) addresses machinery safety and hygiene regulations. Some of the highlights are:

- Machinery containing materials intended to come in contact with food must be designed and constructed so these materials can be cleaned before each use.

- All surfaces and their joinings must be smooth, with no ridges or crevices that could harbor organic materials.

- Assemblies must be designed to minimize projections, edges and recesses. They should be constructed by welding or continuous bonding, with screws, screwheads and rivets used only where technically unavoidable.

- Contact surfaces must able to be readily cleaned and disinfected, and built with easily dismantled parts. Inside surfaces must be curved in a way to allow thorough cleaning.

- Liquid derived from foods, as well as cleaning, disinfecting and rinsing fluids should be able to be readily discharged from machinery.

- Machinery must be designed and constructed to prevent liquids or living creatures - primarily insects - from entering and accumulating in areas that cannot be cleaned.

- Machinery must be designed and constructed so that ancillary substances, such as lubricants, do not come in contact with food.

The directive also sets out a certification system where machinery is checked for compliance and tagged with an EC mark if found to be satisfactory. Certification is not retrospective and manufacturers have two years to bring new machinery into compliance.

Apart from literature already cited, additional useful material and information on hygienic design are found in Anon. (1982, 1983), Milledge (1981) and Katsuyama and Strachan (1980).

7.4. PROCESSING PROCEDURES

Processing procedures are Critical Control Points (CCP-2) in the processing of all food products. All processing techniques and procedures therefore must be designed and aimed at management of contamination and/or growth of microorganisms in food. Such procedures are termed "Good Manufacturing Practices" (GMP).

Detailed codes for GMP must be elaborated for each factory and each processing line (like the HACCP-concept). However a number of details to be included in the GMP-codes have been elaborated by regulatory agencies and international organizations. The most comprehensive example is the work undertaken by the "Codex Alimentarius Commission" of the United Nations, who has published a series of Recommended Codes of Practices (Codex Alimentarius 1969 -) including general principles of food hygiene (Vol. A) and a number of fish products (Vol. B) including codes for fresh fish, canned fish, frozen fish, shrimp, molluscan shellfish, lobsters, crabs, smoked fish, salted fish and minced fish. These codes are continuously updated and should be consulted for detailed information on recommended processing procedures.

7.5. PERSONAL HYGIENE

Personal hygiene is a CCP-2 in preventing microbial contamination or any foreign body contamination of fish products. A list of 15 basic points related to personal hygiene has been drawn up by Thorpe (1992) and are shown below:

Personal hygiene requirements for personnel working in production areas and materials warehouses

1. **Protective clothing, footwear and headgear** issued by the company must be worn and must be changed regularly. When considered appropriate by management, a fine hairnet must be worn in addition to the protective headgear provided. Hair clips and grips should not be worn. Visitors and contractors must comply with this regulation.

2. **Protective clothing** must not be worn off the site and must be kept in good condition. If it is in poor condition, inform your supervisor immediately.

3. **Beards** must be kept short and trimmed and a protective cover worn when considered appropriate by management.

4. **Nail varnish, false nails and make up** must not be worn in production areas.

5. **False eyelashes, wrist watches and jewellery** (except wedding rings, or the national equivalent, and sleeper earrings) must not be worn.

6. **Hands** must be washed regularly and kept clean at all times.

7. **Personal items** must not be taken into production areas unless carried in inside overall pockets (handbags, shopping bags must be left in the locker provided).

8. **Food and drink** must not be taken into or consumed in areas other than the tea bars and the staff restaurant.

9. **Sweets and chewing gum** must not be consumed in production areas.

10. **Smoking or taking snuff** is forbidden in food production, warehouse and distribution areas where 'No Smoking' notices are displayed.

11. **Spitting** is forbidden in all areas on the site.

12. **Superficial injuries** (e.g. cuts, grazes, boils, sores and skin infections) must be reported to the medical department or the first aider on duty via your supervisor and clearance obtained before entering production areas.

13. **Dressings** must be waterproof and contain a metal strip as approved by the medical department

14. **Infectious diseases** (including stomach disorders, diarrhoea, skin conditions and discharge from eyes, nose or ears) must be reported to the medical department or first aider on duty via your supervisor. This also applies to staff returning from foreign travel where there has been a risk of infection.

15. **All staff must report to medical department when returning from both certified and uncertified sickness.**

7.6. APPLICATION OF THE HACCP-PRINCIPLE IN ASSESSMENT OF ESTABLISHMENTS

A great variability exists in the size and extent of handling in fish processing establishments. Accordingly the hygienic requirements and the design in fish handling areas may vary considerably. Quite obviously the requirements to a small establishment which is only repacking fish in ice and catering for a local market are different from the hygienic requirements to a large establishment, processing a variety of sophisticated products including heat treated and composite products and exporting to countries all over the world. However, all the requirements commonly listed in legislation and codes of practice are not equally important. The more important factors include: facilities for water supply, waste disposal and cooling and cold storage facilities and -capacity. Of less importance are buildings, ventilation, factory location, clothes changing facilities, lightning and roadways (ICMSF 1988).

The forms shown in Figure 7.3 have been utilized in assessing fish factories using the HACCP-principle. Only the most important factors are evaluated and given a rating from A to C, where A and B are expressions of degrees of excellence and niceties, while a rating of C is given to a condition which is unacceptable and needs immediate correction before further operations can take place. Thus it is an attempt to "distinguish between the nice and the necessary" which is the same approach as applied in the HACCP-principle.

ASSESSMENT OF FISH FACTORY

Name of factory ____________________ Type of production ________
Name of assessor ____________________ Date for visit ________

FIXED INSTALLATIONS	Rating A	B	C
FACTORY Site (tidiness, pollution) General design, lay-out, flow of goods Separation between clean/unclean processing areas Easy to keep clean Maintenance			
EQUIPMENT Sanitary installations and amenities (toilets, handwashing facilities etc.). Numbers, construction, position Laboratory facilities Water supply (quantity, quality (safe), hot, cold) chlorination Boxes and containers Machinery Waste disposal			
CHILLING/FREEZING CAPACITY Ice supply Chill room (numbers, size/capacity) Freezers/frozen storage (numbers/size/capacity)			
OTHER REMARKS			

VARIABLE FACTORS	A	B	C
RAW MATERIAL quality, handling, control with -			
PROCESS/PROCESS-CONTROL Flow, markings Temperature/temperature control Work routines (GMP/BMP), general tidiness Process control, delegation of responsibility			
PERSONAL HYGIENE Dress General understanding of hygiene principles			
CLEANING AND DISINFECTION Organisation of routine Methods Control with -			
QUALITY ASSURANCE Principles, organisation, delegation of responsibility Staff Monitoring of CCP's, records Procedures for out of control situations			
OTHER REMARKS			

A) Excellent, good or only minor deficiency
B) Less good, serious deficiencies
C) An unacceptable situation, which may result in an unwholesome product representing health of safety threats.

Figure 7.3. Example of simplified form used in assessing fish factories.

8. REFERENCES

Ababouch, L., M.E. Afilal, H. Benabdeljelil and F.F. Busta 1991. Quantitative changes in bacteria, amino acids and biogenic amines in sardines (*Sardina pilchardus*) stored at ambient temperature (25-28°C) and in ice. *Int. J. Food Sci. Technol.* **26**, 297-306.

Ackman, R.G. and W.M.N. Ratnayake 1992. Non enzymatic oxidation of seafood lipids. In *Advances in Seafood Biochemistry*. Eds: G.J. Flick Jr. and R. Martin. Technomic Publishing Co., Basel, 245-267.

Addison, R.F. and J.E. Stewart 1989. Domoic acid and the Eastern Canadian molluscan shellfish industry. *Aquaculture* **77**, 263-269.

Ahmed, F.E. (Ed.) 1991. *Seafood Safety*. National Academy Press, Washington D.C., USA.

Angelotti, R. 1970. The heat resistance of *C.botulinum* type E in Food. In *Toxic Microorganisms*. Ed: M. Herzberg. US Dept. of the Interior, 404-409.

Anon. 1972. Proceedings of the 1971 National Conference on Food Protection. US Government Printing Office, Washington DC, USA.

Anon. 1973. The Double Seam Manual. The Metal Box Co. plc: Reading.

Anon. 1982. R.A. Campden. Technical Memorandum 289. *The Principles of Design for Hygienic Food Processing Machinery*. Campden Food Preservation Research Association, Chipping Campden, Gloucestershire GL55 6LD, United Kingdom.

Anon. 1983. R.A. Campden. Technical Manual No. 7. *Hygienic Design of Food Processing Equipment*. Campden Food Preservation Research Association, Chipping Campden, Gloucestershire GL55 6LD, United Kingdom.

AOAC/FDA (Association of Official Analytical Chemists/United States Food and Drug Administration) 1984. *Classification of Visible Can Defects (exterior)*. Ass. of Analytical Chemists, Arlington.

Arai, T., N. Jkejima and T. Itoh 1980. A survey of *Plesiomonas shigelloides* from aquatic environments, domestic animals, pets and humans. *J. Hyg.* **84**, 203-211.

Archer, D.L. 1992. Policy on *Listeria* in Food: An FDA Perspective. In *A Communication* (Book of Abstracts) at the 11th International *Symposium on Problems of Listeriosis*. ISOPOL XI, 11-14 May, Statens Seruminstitut, Copenhagen, Denmark. Abstract 66.

Barile, L.E., M.H. Estrada, A.D. Milla, A. Reilly and A. Villadsen 1985. Spoilage patterns of mackerel (*Rastrelliger faughni* Matsui) 2. Mesophilic and psychrophilic spoilage. *ASEAN Food. J.* **1**, 121-126.

Bauman, H.E. 1992. Introduction to HACCP. In *HACCP Principles and Applications*. Eds: M.D. Pierson and D.A. Corlett, Jr. Van Nostrand Reinhold, 1-5.

Bean, N.H. and P.M. Griffin 1990. Foodborne disease outbreaks in the United States 1973-1987. Pathogens, vehicles, and trends. *J. Food Protect.* **53**, 804-817.

Beckers, H.J. 1986. Incidence of Foodborne Diseases in The Netherlands: Annual Summary, 1981. *J. Food Protect.* **53**, 924-931.

Ben Embarek, P.K. and H.H. Huss 1992. Growth of *Listeria monocytogenes* in lightly preserved fish products. In *Quality Assurance in the Fish Industry*. Eds: H.H. Huss, M. Jacobsen and J. Liston. Elsevier Science Publishers, 293-304.

Ben Embarek, P.K. and H.H. Huss 1993. Heat resistance of *Listeria monocytogenes* in vacuum packaged pasteurized fish fillets. *Int. J. Food Microbiol.* In press.

Bille, J., D. Nocera, E. Bannerman and F. Ischer 1992. Molecular typing of *Listeria monocytogenes* in relation with the Swiss outbreak of listeriosis. Proceedings of the 11th *International Symposium on Problems of Listeriosis*. ISOPOL XI, 11-14 May, Statens Seruminstitut, Copenhagen, Denmark, 195-196.

Blake, P.A., D.T. Allegra and J.D. Snyder 1980. Cholera - a possible epidemic focus in the U.S. *New Eng. J. Med.* **302**, 305-309.

Bradshaw, J.G., D.B. Shah, A.J. Wehby, J.T. Peeler and R.M. Twedt 1984. Thermal inactivation of the Kanagawa hemolysin of *Vibrio prarahaemolyticus* in buffer and shrimp. *J. Food Sci.* **49**, 183-187.

Bremner, H.A., J. Olley and A.M.A. Vail 1987. Estimating time-temperature effects by a rapid systematic sensory method. In *Seafood Quality Determination*. Eds: D. Kramer and J. Liston. Elsevier Science Publishers, 413-435.

Brier, J.W. 1992. Emerging problems in seafood-borne parasitic zoonoses. *Food Control* **3**, 2-7.

Bryan, F.L. 1980. Epidemiology of foodborne diseases transmitted by fish, shellfish and marine crustaceans in the United States, 1970-1978. *J. Food Protect.* **43**, 859-876.

Bryan, F.L. 1987. Seafood-transmitted infections and intoxications in recent years. In *Seafood Quality Determination*. Eds: D.E. Kramer and J. Liston. Elsevier Science Publishers, 319-337.

Bryan, F.L. 1988. Risks associated with vehicles of foodborne pathogens and toxins. *J. Food Prot.* **51**, 498-508

Buckle, K.A. (Ed.) 1989. *Foodborne Microorganisms of Public Health Significance.* AIFST (NSW Branch), Food Microbiology Group, P.O.Box 277, Pymble, NSW 2073, Australia.

Cann, D.C. and L.Y. Taylor 1979. The control of the botulism hazard in hot-smoked trout and mackerel. *J. Food Technol.* **14**, 123, 129.

Cliver, D.O. 1988. Virus transmission via foods. *Food Technol.* **42**, 241-248.

Codex Alimentarius 1969. Vol. A (*Recommended code of practice, general principles of food hygiene*); Vol. B (*Recommended code of practices for fish and fish products*). FAO Documents Office. Joint FAO/WHO Food Standards Programme, Via delle Terme di Caracalla, 00100 Rome, Italy.

Colwell, R.R. 1986. *Vibrio cholerae* and related vibrios in the aquatic environment - an ecological paradigm. *J. Appl. Bacteriol.* **61**, vii.

Condon, S., M.L. Garcia, A. Otero and F.J. Sala 1992. Effect of culture age, pre-incubation at low temperature and pH on the thermal resistance of *Aeromonas hydrophila. J. Appl. Bacteriol.* **72**, 322-326.

Connor, D.E., V.N. Scott and D.T. Bernard 1989. Potential *Clostridium botulinum* hazards associated with extended shelf-life refrigerated foods: A review. *J. Food Safety* **10**, 131-153

Dainty, R.H., B.G. Shaw and T.A. Roberts 1983. Microbial and chemical changes in chill-stored red meats. In *Food Microbiology. Advances and Prospects*. Eds: T.A. Roberts and F.A. Skinner. Academic Press, 151-178.

Dalgaard, P., L. Gram and H.H. Huss 1993. Spoilage and shelf life of cod fillets packed in vacuum or modified atmosphere. *Int. J. Food Microbiol.* **19**, 283-294.

D'Aoust, J.Y,, R. Gelinas and C. Maishment 1980. Presence of indicator organisms and recovery of *Salmonella* in fish and shellfish. *J. Food Protect.* **43**, 769-782.

D'Aoust, J.Y. 1989. *Salmonella.* In *Foodborne Bacterial Pathogens*. Ed: M.P. Doyle. Marcel Dekker Inc., 327-445.

Delmore, R. and P. Crisley 1979. Thermal resistance of *V. parahaemolyticus* in a clam homogenate. *J. Food Sci.* **41**, 899-902.

Donald, B. and D. Gibson 1992. *Spoilage of MAP salmon steaks stored at 5°C.* EEC report on the FAR project UP-2-545. Torry Research Station, Aberdeen, Scotland.

Doyle, M.P. (Ed.) 1989. *Foodborne Bacterial Pathogens*. Marcel Dekker Inc.

EEC 1980. Council Directive 80/778/EEC of 15 July 1980 relating to the quality of water intended for human consumption. *Official Journal of the European Communities* No. L 229, 30.08.1980, 11.

EEC 1989. Council Directive 89/392/EEC of 14 June 1989 on the approximation of the laws of the member states relating to machinery. *Official Journal of the European Communities* No. L 183, 29.06.1989, 9-32.

EEC 1991a. Council Directive 91/492/EEC of 15 July 1991 laying down the health conditions for the production and the placing on the market of live bivalve molluscs. *Official Journal of the European Communities* No. L 268, 24.09.1991, 1.

EEC 1991b. Council Directive 91/493/EEC of 22 July laying down the health conditions for the production and the placing on the market of fishery products. *Official Journal of the European Communities* No. L 268, 24.09.1991, 15.

EEC 1992. Proposal for a Council Directive on the hygiene of foodstuffs. *Official Journal of the European Communities* No. C 24/11, 31.01.1992, 11-16.

Eyles, M.J. 1986. Transmission of viral disease by food: an update. *Food Technol. Aust.* **38**, 239-242.

Eyles, M.J. 1989. Viruses. In *Foodborne Microorganisms of Public Health Significance*. 4th ed. Ed: Buckle, K.A. AIFST (NSW Branch) Food Microbiology Group. P.O. Box 277, Pymble, NSW 2073, Australia.

Facinelli, B., P.E. Varaldo, C. Casolari and V. Fabio 1988. Cross-infection with *Listeria monocytogenes* confirmed with DNA-finger printing. *Lancet* II (8622), 1247-1248.

Facinelli, B., P.E. Varaldo, M. Toni, C. Casolari and V. Fabio 1989. Ignorance about *Listeria*. *Bri. Med. J.* **299**, 738.

FAO/WHO 1979. *Recommended International Code of Practice for Low-acid and Acidified Canned Foods* (CAC/RCP 23-1979).

FAO 1989. *Food safety regulations applied to fish by the major importing countries*. FAO Fisheries Circular No. 825, FAO, Rome.

Farber, I.M. 1986. Predictive modeling of food deterioration and safety. In *Foodborne Microorganisms and their Toxins: Developing Methodology*. Eds: M.D. Person and N.J. Sterns. Marcel Dekker Inc., 57-90.

Farber, J.M. and P.J. Peterkin 1991. *Listeria monocytogenes*, a food-borne pathogen. *Microbiol. Rev.* **55**, 476-511.

FDA (Food and Drug Administration) 1973. Thermally processed low-acid foods packed in hermetically sealed containers. Part 128B (recodified as part 113). *Federal Register*, January **38**, 2398-2410.

FDA (Food and Drug Administration) 1989. *National Shellfish Sanitation Program. Manual of Operations.* Center for Food Safety and Applied Nutrition, Division of Cooperative Programs, Shellfish Sanitation Branch, Washington D.C., USA.

Fisheries and Oceans 1983. *Identification and Classification Manual.* Canadian Department of Fisheries and Oceans, Ottawa, Canada.

FMF/FMA 1967. Joint Technical Committee, Food Manufacturers Federation (FMF) and Food Machinery Association (FMA): Hygienic Design of Food Plant London, United Kingdom.

FNB/NRC (Food and Nutrition Board, National Research Council, USA) 1985. *An evaluation of the role of microbiological criteria for foods and food ingredients* (Sub-committee on Microbiological Criteria, Committee on Food Protection). National Academy Press, Washington D.C., USA.

Food Safety Act 1990. Chapter 16, HMSO, London, United Kingdom.

Frederiksen, W. 1991. *Listeria* epidemiology in Denmark 1981-1990. In *Proceedings of Int. Conference on Listeria and Food Safety*, 13-14 June. Laval, France. ASEPT Eds., 48-49.

Fuchs, R.S. and P.J.A. Reilly 1992. The incidence and significance of *Listeria monocytogenes* in seafoods. In *Quality Assurance in the Fish Industry*. Eds: H.H. Huss, M. Jakobsen and J. Liston. Elsevier Science Publishers, 217-230.

Fujioka, R.S., K. Tenno and S. Kansako 1988. Naturally occurring fecal coliforms and fecal streptococci in Hawaii's freshwater streams. *Toxic Assess.* **3**, 613-630.

Garrett, E.S. and M. Hudak-Roos 1991. Developing an HACCP-Based inspection system for the seafood industry. *Food Technol.* **45**, 53-57.

Gerba, C.P. 1988. Viral disease transmission by seafoods. *Food Technol.* **42**, 99-102.

Gerba, C.P. and S.M. Goyal 1978. Detection and occurrence of enteric viruses in shellfish: A review. *J. Food Protect.* **41**, 742.

German Fish Ordinance 1988. Bundesminister für Jugend, Familie, Frauen und Gesundheit 1988. Verordnung über gesundheitliche auforderungen an Fische- und Schalentiere (Fisch-Verordnung). *Bundesgesetzblatt*, 1570.

Gerner-Smit, P. and B. Nørrung 1992. Comparison of four different typing methods for *Listeria monocytogenes* using a newly described discriminatory index. Proceedings of the 11th *International Symposium on Problems of Listeriosis*. ISOPOL XI, 11-14 May, Statens Seruminstitut, Copenhagen, Denmark, 199-200.

Gibson, D. 1992. *Personal communication.* Torry Research Station. Aberdeen Scotland.

Gill, T.A., J.W. Thompson and S. Gould 1985. Thermal Resistance of Paralytic Shellfish Poison in Soft Shell Clams. *J. Food Protect.* **48**, 659-662.

Gorczyca, E. and Pek Poh Len 1985. Mesophilic spoilage of bay trout (*Arripis trutta*), bream (*Acanthropagrus butcheri*) and mullet (*Aldrichetta forsteri*). In *Spoilage of tropical fish and product development*. Proceedings of a symposium held in conjunction with the Sixth Session of The Indo-Pacific Fishery Commission Working Party on Fish Technology and Marketing. Ed: A. Reilly. Royal Melbourne Institute of Technology, Melbourne, Australia 23-26 October, 1984. FAO Fish Rep. (317) Suppl. 123-132.

Gram, L., G. Trolle and H.H. Huss 1987. Detection of specific spoilage bacteria on fish stored at high (20°C) and low (0°C) temperatures. *Int. J. Food Microbiol.* **4**, 65-72.

Gram, L., C. Wedell-Neergaard and H.H. Huss 1990. The bacteriology of fresh and spoiling Lake Victorian Nile perch (*Lates niloticus*). *Int. J. Food Microbiol.* **10**, 303-316.

Guerrant, R.L. 1985. Microbial toxins and diarrhoeal diseases: Introduction and Overview. In *Microbial toxins and diarrhoeal diseases*. Eds: E. Evered and J. Wheland. CIBA Foundation Symposium No. 112. Pitman Publishing London.

Guyer, S. and T. Jemmi 1991. Behavior of *Listeria monocytogenes* during fabrication and storage of experimentally contaminated smoked salmon. *Appl. Environ. Microbiol.* **57**, 1523-1527.

Hall, S. 1991. Natural toxins. In *Microbiology of Marine Food Products*. Eds: D.R. Ward and C. Hackney. Van Nostrand Reinhold, 301-330.

Halstead, B.W. 1978. *Poisonous and venomous marine animals of the world*. Princeton Darwin Press.

Halstead, B.W. and E.J. Schantz 1984. *Paralytic Shellfish Poisoning*. WHO Offset Publication No. 79, Geneva.

Harrigan, W.F. 1993. The ISO 9000 series and its implications for HACCP. *Food Control* **4**, 105-111.

Hauschild, A.N.W. 1989. *Clostridium botulinum*. In *Foodborne Bacterial Pathogens*. Ed: M.P. Doyle. Marcel Dekker Inc., 111-189.

Hayes, P.R. 1985. *Food Microbiology and Hygiene*. Elsevier Applied Science.

Hazen, T. 1988. Fecal coliforms as indicators in tropical waters: A review. *Toxic Assess.* **3**, 461-477.

Healy, G.R. and D. Juranek 1979. Parasitic infections. In *Food-Borne Infections and Toxications*. Eds: H. Riemann and F.L. Bryan. Academic Press, 343-385.

Hersom, A.C. and E.D. Hulland 1980. *Canned Foods. Thermal Processing and Microbiology*. 7. ed. Churchill Livingstone, Edinburgh, 380.

Herrington, D.A., S. Tzipori, R.M. Robins-Browne, B.D. Tall and M.M. Levine 1987. In vitro and in vivo pathogenity of *Plesiomonas shigelloides*. *Infect. Immun.* **55**, 979-985.

Higashi, G.J. 1985. Foodborne parasites transmitted to man from fish and other aquatic foods. *Food Technol.* **39**, 69.

Hudak-Roos, M. and E.S. Garrett 1992. Monitoring critical control points critical limits. In *HACCP Principles and Applications*. Eds: M.D. Pierson and D.A. Corlett, Jr. Van Nostrand Reinhold, 62-71.

Huss, H.H. 1980. Distribution of *Clostridium botulinum*. *Appl. Environ. Microbiol.* **39**, 764-769.

Huss, H.H. 1981. *Clostridium botulinum type E and botulism.* Thesis. Technological Laboratory, Ministry of Fisheries, Technical University, Lyngby, Denmark.

Huss, H.H. 1988. *Fresh fish. Quality and Quality Changes*. FAO Fisheries Series No. 29.

Huss, H.H. and A. Pedersen 1979. *Clostridium botulinum* in fish. *Nord. Vet. Med.* **31**, 214-221.

Huss, H.H. and E. Rye Petersen 1980. The stability of *Clostridium botulinum* type E toxin in salty and/or acid environment. *J. Food Technol.* **15**, 619-627.

Huss, H.H., D. Dalsgaard, L. Hansen, H. Ladefoged, A. Pedersen and L. Zittan 1974. The influence of hygiene in catch handling on the storage life of iced cod and plaice. *J. Food Technol.* **9**, 213-221.

Huss, H.H., A. Roepstorff, H. Karl and B. Bloemsma 1992. Handling and processing of herring infected with *Anisakis simplex*. In *Proceedings from 3rd World Congress on Foodborne Infections and Intoxications*. Inst. of Vet. Med. Robert v. Ostertag-Inst. Berlin, 388-394.

IAMFES 1991. *Procedures to implement the Hazard Analysis Critical Control Point System.* Int. Ass. Milk, Food and Environ. Sanitarians Inc. 502. E. Lincoln Way. Ames, Iowa, 50010-6666 U.S.A.

ICMSF (International Commission on Microbial Specifications for Foods) 1986. *Microorganisms in Foods. 2. Sampling for microbiological analysis: Principles and specific applications.* 2nd ed. Blackwell Scientific Publications.

ICMSF (International Commission on Microbial Specifications for Foods) 1988. *Microorganisms in foods. 4. Application of the Hazard Analysis Critical Control Point (HACCP) system to ensure microbiological safety and quality.* Blackwell Scientific Publications.

Imholte, T.J. 1984. *Engineering for Food Safety and Sanitation.* Crystal, MINN: The Technical Institute of Food Safety.

ISO 8402. *Quality - Vocabulary*

ISO 9000. *Quality management and quality assurance standards - Guidelines for selection and use.*

ISO 9001. *Quality systems - Model for quality assurance in design/development, production, installation and servicing.*

ISO 9002. *Quality systems - Model for quality assurance in production and installation.*

ISO 9003. *Quality systems - Model for quality assurance in final inspection and test.*

ISO 9004. *Quality management and quality system elements - Guidelines.*

Jiménez, L., J. Munir, G.G. Toranzos and T.C. Hazen 1989. Survival and activity of *Salmonella typhimurium* and *Escherichia coli* in tropical fresh water. *J. Appl. Bacteriol.* **67**, 61-69.

Jørgensen, B.R., D.M. Gibson and H.H. Huss 1988. Microbiological quality and shelf life prediction of chilled fish. *Int. J. Food Microbiol.* **6**, 295-307.

Jørgensen, B.R. and H.H. Huss 1989. Growth and activity of *Shewanella putrefaciens* isolated from spoiling fish. *Int. J. Food Microbiol.* **9**, 51-62.

Karunasagar, I., K. Segar, I. Karunasagar and W. Goebel 1992. Incidence of *Listeria* spp. in Tropical Seafoods. In Proceedings of the 11th *International Symposium on Problems of Listeriosis.* ISOPOL XI, 11-14 May, Statens Seruminstitut, Copenhagen, Denmark, 306.

Katsuyama, A.M. and J.P. Strachan (Eds.) 1980. *Principles of Food Processing Sanitation.* The Food Processors Institute. Washington, DC, 303.

Kilgen, M.B. and M.T. Cole 1991. Viruses in seafood. In *Microbiology of Marine Food Products.* Eds: D.R. Ward and C. Hackney. Van Nostrand Reinhold, 197-209.

Klausen, N.K. and H.H. Huss 1987. Growth and histamine production by *Morganella morganii* under various temperature conditions. *Int. J. Food Microbiol.* **5**, 147-156.

Knøchel, S. 1989. *Aeromonas spp. - Ecology and significance in food and water hygiene.* Ph.D. Thesis. The Royal Veterinary and Agricultural University, Copenhagen, Denmark.

Knøchel, S. 1990. *Microbiology and Groundwater: Aesthetic and Hygienic Problems.* Water Quality Institute, Hørsholm, Denmark.

Knøchel, S. and H.H. Huss 1984. Ripening and spoilage of sugar salted herring with and without nitrate. I. Microbiological and related chemical changes. *J. Food Technol.* **19**, 203-213, 215-224.

Koburger, J.A. 1989. *Plesiomonas shigelloides.* In *Foodborne Bacterial Pathogens.* Ed: M.P. Doyle. Marcel Dekker Inc., 311-325.

Lennon, D., B. Lewis, C. Mantell, D. Becroft, B. Dove, K. Farmer, S. Tonkin, N. Yeats, R. Stamp and K. Mickleson 1984. Epidemic perinatal listeriosis. *Pediatr. Infect. Dis.* **3**, 30-34.

Lewis, K.H. 1980. Cleaning, disinfection & hygiene. In *Microbial Ecology of Foods. Vol. 1: Factors Affecting Life and Death of Microorganisms.* International Commission on Microbiological Specifications for Foods, Academic Press.

Lima dos Santos, C.A.M. 1978. *Bacteriological Spoilage of Iced Amazonian Freshwater Catfish (Brachyplatistoma vaillanti valenciennes).* Master's Thesis Loughborough University of Technology.

Lovett, J. 1989. *Listeria monocytogenes.* In *Foodborne Bacterial Pathogens.* Ed: M.P. Doyle. Marcel Dekker Inc., 283-310.

Lupin, H. 1992. *Personal communication.* FAO Rome, Italy.

Mackey, B.M. and N. Bratchell 1989. The heat resistence of *Listeria monocytogenes.* A review. *Lett. Appl. Microbiol.* **9**, 89-94.

Marshall, B.J., D.F. Ohye and J.H.B. Christian 1971. Tolerance of bacteria to high concentrations of NaCl and glycerol in the growth medium. *Appl. Microbiol.* **21**, 363-364.

Matsui, T., S. Taketsuyu, K. Kodama, A. Ishii, K. Yamamori and C. Shimizu 1989. Production of tetrodotoxin by the intestinal bacteria of a pufferfish *Takifugu niphobes. Nippon Suisan Gokkaishi* **55**, 2199-2203.

Mayes, T. 1992. Simple users' guide to the Hazard Analysis Critical Control Point concept for the control of microbiological safety. *Food Control* **3**, 14-19.

Melnick, J.C. and C.P. Gerba 1980. The ecology of enteroviruses in natural waters. *CRC Crit. Rev. Environ.* **10**, 65.

Milledge, J.J. 1981. The hygienic design of food plant. *Institute of Food Science and Technology (U.K.). Proceedings* **14**, 74-86.

Miller, M.L. and J.A. Koburger 1986. Tolerance of *Plesiomonas shigelloides* to pH, sodium chloride and temperature. *J. Food Prot.* **49**, 877-879.

Miller, S.A. and J.E. Kvenberg 1986. Reflections on food safety. Presented at the National Food Processors Association (NFPA) Conference Avoiding Pesticide, Environmental Contaminants and Food Additives Crisis. Atlanta GA, 3 February. NFPA Washington D.C., USA.

Mitchell, B. 1992. How to HACCP. *British Food J.* **94**, 1, 16-20.

Mitscherlich, E. and E.H. Marth 1984. *Microbial Survival in the Environment.* Springer-Verlag.

Morris, J.G. and R.E. Black 1985. Cholera and other vibrios in the United States. *N. Engl. J. Med.* **312**, 343-350.

Mossel, D.A.A. 1967. Ecological principles and methodological aspects of the examination of foods and feeds for indicator microorganisms. *J. Assoc. Agric. Chem.* **50**, 91-104.

Mossel, D.A.A. 1982. *Microbiology of Foods.* University of Utrecht. Faculty of Vet. Med., Bittshact 172, Utrecht, The Netherlands.

Mossel, D.A.A. and D.M. Drake 1990. Processing food for safety and reassuring the consumer. *Food Technol.*

Mossel, D.A.A., A. Veldman and J. Eeldering 1980. Comparison of the effects of liquid medium repair and the incorporation of catalase in Mac Conkey type media on the recovery of *Enterobacteriaceae* sublethally stressed by freezing. *J. Appl. Bacteriol.* **49**, 405-419.

NACMCF (U.S. National Advisory Committee on Microbial Criteria for Foods) 1992. Hazard Analysis Critical Control Point System. *Int. J. Food Microbiol.* **16**, 1-23.

Noguchi, T., D.F. Hwang, O. Arakawa, H. Sugita, Y. Deguchi, Y. Shida and K. Hashimoto 1987. *Vibrio alginolyticus*, a tetrodotoxin producing bacterium in the intestines of the fish *Fugu vermiculans vermi cularis. Mar. Biol.* **94**, 625.

Nolan, D.A., D.C. Chamblin and J.A. Troller 1992. Minimal water activity levels for growth and survival of *Listeria monocytogenes* and *Listeria innocua. Int. J. Food Microbiol.* **16**, 323-335.

Notermans, S., S.R. Tafini and T. Chakraborty 1992. An approach to set realistic criteria for *Listeria* in food products. In *A Communication* (Book of Abstracts) at the 11th *International Symposium on Problems of Listeriosis*. ISOPOL, 11-14 May, Statens Seruminstitut, Copenhagen, Denmark. Abstract 65.

Olson, R.E. 1987. Marine fish parasites of public health importance. In *Seafood Quality Determination*. Eds: D.E. Kramer and J. Liston. Elsevier Science Publishers, 339-355.

Pace, P.J. and E.R. Krumbiegel 1973. *Clostridium botulinum* and smoked fish production 1963-72. *J. Milk Food Technol.* **36**, 42-49.

Palumbo, S.A., D.R. Morgan and R.L. Buchanan 1985. Influence of temperature, NaCl and pH on growth of *Aeromonas hydrophila*. *J. Food Sci.* **50**, 1417-1421.

Pan, B.S. and D. James (Eds.) 1985. *Histamine in marine products: production by bacteria, measurement and prediction of formation.* FAO Fish. Tech. Paper (252).

Pierson, M. and D.A. Corlett Jr. 1992. *Appendix B in HACCP. Principles and Applications.* van Nostrand Reinhold.

Poretti, M. 1990. Quality control of water as raw material in the food industry. *Food Control* **1** (2), 79-83.

Prasad, V.S. and M. Chaudhuri 1989. Development of filtration/adsorption media for removal of bacteria and turbidity from water. *Wat. Sci. Tech.* **21**, 67-71.

Premazzi, G., G. Chiaudani and G. Ziglio 1989. *Scientific Assessment of EC Standards for Drinking Water Quality.* European Communities Commission, Luxembourg.

Price, R.J. 1992. Residue concerns in seafoods. *Dairy, food and Environmental Sanitation* **12**, 139-143

Ragelis, E.P. 1984. Ciguatera seafood poisoning. Overview. In *Seafood Toxins*. Ed: E.P. Ragelis. ACS Symposium Series 262. Washington D.C., 25-36.

Reilly, P.J.A., D.R. Twiddy and R.S. Fuchs 1992. *Review of the occurrence of salmonella in cultured tropical shrimp.* FAO Fisheries Circular No. 851. FAO, Rome.

Rhodes, M.W. and H. Kator 1988. Survival of *Escherichia coli* and *Salmonella* spp. in estaurine environments. *Appl. Environ. Microbiol.* **54**, 2902-2907.

Richards, G.P. 1985. Outbreaks of shellfish associated enteric virus illness in the United States: requisite for development of viral guidelines. *J. Food Prot.* **48**, 815-823.

Richards, G.P. 1991. Shellfish depuration. In *Microbiology of Marine Food Products*. Eds: D.R. Ward and C.R. Hackney. Van Nostrand Reinhold, 395-428.

Riedo, F.X., R.W. Pinner, M. Tosca, M.L. Carter, L.M. Graves, M.W. Reaves, B.D. Plikaytis and C.V. Broome 1990. Program Abstracts: 30th *Intersci. Conference on Antimicrobial Agents and Chemotherapy*, abstr. 972.

Rim, H.J. 1982. Clonorchiasis. In *CRC Handbook Series in Zoonosis, Section C: Parasitic Zoonoses*. Eds: G.V. Hillyer and C.F. Hopla. Vol. III, CRC Press Inc., Boca Roton, FL., 17.

Roos, R. 1956. Hepatitis epidemic conveyed by oysters. *Svenska Läkartidningen* **53**, 989.

Ryser, E.T. and E.H. Marth 1991. *Listeria, Listeriosis and Food Safety*. Marcel Dekker Inc.

Rørvik, L.M. and M. Yndestad 1991. *Listeria monocytogenes* in Foods in Norway. *Int. J. Food Microbiol.* **13**, 97-104.

Rørvik, L.M., M. Yndestad and E. Skjerve 1991. Growth of *Listeria monocytogenes* in vacuum packed smoked salmon during storage at 4°C. *Int. J. Food Microbiol.* **14**, 111-117.

Sattler, J. and W. Lorenz 1990. Intestinal diamine oxidases and enteral-induced histaminosis: studies of three prognostic variables in an epidemiological model. *J. Neural Transm.* Suppl. **32** 291-314

Schantz, E.J. 1984. Historical perspective on paralytical shellfish poisoning. In: *Seafood Toxins*. Ed: E.P. Ragelis. ACS - Symposium Series 262, 99-111.

Shahamat, M., A. Seaman and M. Woodbine 1980. Influence of sodium chloride, pH and temperature on the inhibiting activity of nitrite on *Listeria monocytogenes*. In: *Survival in Extremes of Environment*. Eds: G.W. Gould and J.E.L. Corry. Academic Press.

Shapton, D.A. and N.F. Shapton 1991. *Principles and Practices for the Safe Processing of Food.* Butterworth & Heinemann.

Shultz, L., J. Rutledge, R. Grudner and S. Biede 1984. Determination of thermal death time and *V. cholerae* in blue crab. *J. Food Protect.* **47**, 4.

Silliker, J.H. and D.A. Gabis 1976. ICMSF method studies VII. Indicator tests as substitutes for direct testing of dried foods and feeds for *Salmonella*. *Can. J. Microbiol.* **22**, 971-974.

Skovgaard, N. 1992. *Listeria monocytogenes* in raw food - an overview. In *A Communication* (Book of Abstracts) at the 11th *International Symposium on Problems of Listeriosis*. ISOPOL, 11-14 May, Statens Seruminstitut, Copenhagen, Denmark. Abstract 64.

Sobsey, M.D. 1989. Inactivation of health-related microorganisms in water by disinfection processes. *Wat. Sci. Tech.* **21** (3), 179-195.

Stratten, J.E. and S.L. Taylor 1991. Scombroid poisoning. In *Microbiology of Marine Food Products*. Eds: D.R. Ward and C.R. Hackney. Van Nostrand Reinhold, 331-351.

Tang, Y.W., J.X. Wang, Z.Y. Xu, Y.F. Guo, W.H. Qian and J.x. Xu 1991. A serologically confirmed case-control study of a large outbreak of hepatitis A in China, associated with consumption of clams. *Epidemiol. Infect.* **107**, 651-657.

Taylor, S.L. 1986. Histamine food poisoning: Toxicology and clinical aspects. *CRC Crit. Rev. Toxicol.* **17**, 91-128.

Taylor, S.L. 1988. Marine toxins of microbial origin. *Food Technol.* **42**, 94-98.

Thorpe, R.H. 1992. Hygienic design considerations for chilled food plants. In *Chilled Foods. A Comprehensive guide.* Eds: C. Dennis and M. Stringer. Ellis Horwood.

Thorpe, R.H. and P.M. Barker 1984. *Visual Can Defects.* The Campden Food Preservation Research Ass. Chipping Campden, United Kingdom.

Todd, E.C.D. 1989a. Foodborne and waterborne disease in Canada - 1983. Annual summary. *J. Food Protect.* **52**, 436-442.

Todd, E.C.D. 1989b. Preliminary estimates of costs of foodborne disease in the United States. *J. Food Protect.* **52**, 595-601.

Todd, E.C.D. 1990. Amnesic Shellfish Poisoning - A new seafood toxic syndrome. In: *Toxic Marine Phytoplancton.* Eds: E. Graneli, B. Sundstrøm, L. Edlar and D.M. Andersen. Elsevier, 504-508.

Todd, E.C.D. 1993. Domoic Acid and Amnesic Shellfish Poisoning. A review. *J. Food Prot.* **56**, 69-86.

Tompkin, R.B. 1990. The use of HACCP in the production of meat and poultry products. *J. Food Prot.* **53**, 795-803.

Tompkin, R.B. 1992. Corrective Action Procedures for deviations from critical control point critical limits. In *HACCP, Principles and Applications*. Eds: M.D. Pierson, D.A. Corlett Jr. Van Nostrand Reinhold, 72-82.

Toranzos, G.A., C.P. Gerba and H. Hansen 1988. Enteric viruses and coliphages in Latin America. *Toxic Assess.* **3**, 491-510.

Troller, J.A. 1983. *Sanitation in Food Processing.* Academic Press.

Turnbull, P.C.B. and R.J. Gilbert 1982. Fish and shellfish poisoning in Britain. In: *Adverse Effects of Foods.* Eds: E.F.P. Jellife and D.B. Jellife. Plenum Press, 297-306.

Valdimarsson, G. 1989. Future aspects of fish processing. In *Nutritional Impact of Food Processing.* Eds. J.C. Somogyi and H.R. Müller. *Bibl. Nutr. Dieta Basel* **43**, 78-88.

Van Spreekens, K.J.A. 1977. Characterization of some fish and shrimp spoiling bacteria. *Ant. van Leeuwenhoek* **43**, 282-303.

Van Spreekens, K.J.A. 1987. Histamine production by the psycrophilic flora. In *Seafood Quality Determinations*. Eds: D. Kramer and J. Liston. Elsevier Science Publishers, 309-318.

Varnam, A.H. and M.G. Evans 1991. *Foodborne Pathogens*. Wolfe Publishing Ltd.

Wachsmuth, K. and G.K. Morris 1989. *Shigella*. In *Foodborne Bacterial Pathogens*. Ed: M.P. Doyle. Marcel Dekker Inc., 447-462.

Walker, S.J. and M.F. Stringer 1987. Growth of *Listeria monocytogenes* and *Aeromonas hydrophila* at chill temperatures. *Tech. Memo.*, **462**, CFPRA Chipping Campden, United Kingdom.

Ward, R.L. and E.W. Akin 1983. Minimum infective dose of animal viruses. *Crit. Rev. Environ. Control* **14**, 297-310.

Weagant, S.D., P.N. Sado, K.G. Colburn 1989. The incidence of *Listeria* species in frozen seafood products. *J.Food Protect.* **51**, 655-657.

White, D.R.L. and J.E.P. Noseworthy 1992. The Canadian Quality management Programme. In: *Quality assurance in the Fish industry*. Eds: H.H. Huss, M. Jakobsen and J. Liston. Elsevier Science Publishers, 509-513.

WHO 1984a. Aquatic (Marine and Freshwater) Biotoxins. *Environ. Health Criter.* **37**, Geneva.

WHO 1984b. *Guidelines for Drinking Water Quality*, Vols. 1, 2, 3. World Health Organisation, Geneva.

WHO 1989. *Report of WHO Consultation on Public Health Aspects of Seafood-Borne Diseases*. WHO/CDS/VPH/90.86.

WHO 1992. *WHO guidance on formulation of national policy on the control of cholera*. WHO ICDD/SER/80.4 Rev. 3.

9. INDEX

E

F

G

H

I

W